S0-AKJ-172

An Introduction to
Laboratory Automation

CHEMICAL ANALYSIS

A SERIES OF MONOGRAPHS ON
ANALYTICAL CHEMISTRY AND ITS APPLICATIONS

Editors
J. D. WINEFORDNER
Editor Emeritus: **I. M. KOLTHOFF**

Advisory Board

Fred W. Billmeyer, Jr.
Eli Grushka
Barry L. Karger
Viliam Krivan

Victor G. Mossotti
A. Lee Smith
Bernard Tremillon
T. S. West

VOLUME 110

A WILEY-INTERSCIENCE PUBLICATION

John Wiley & Sons, Inc.

NEW YORK / CHICHESTER / BRISBANE / TORONTO / SINGAPORE

An Introduction to Laboratory Automation

VÍCTOR CERDÁ

Department of Chemistry
University of the Balearic Islands
Palma de Mallorca, Spain

GUILLERMO RAMIS

Department of Analytical Chemistry
University of Valencia
Burjassot, Valencia, Spain

A WILEY-INTERSCIENCE PUBLICATION

John Wiley & Sons, Inc.

NEW YORK / CHICHESTER / BRISBANE / TORONTO / SINGAPORE

In recognition of the importance of preserving what has
been written, it is a policy of John Wiley & Sons, Inc. to
have books of enduring value published in the United
States printed on acid-free paper, and we exert our best
efforts to that end.

Copyright © 1990 by John Wiley & Sons, Inc.

All rights reserved. Published simultaneously in Canada.

Reproduction or translation of any part of this work
beyond that permitted by Section 107 or 108 of the
1976 United States Copyright Act without the permission
of the copyright owner is unlawful. Requests for
permission or further information should be addressed to
the Permissions Department, John Wiley & Sons, Inc.

Library of Congress Cataloging in Publication Data:

Cerda, Víctor.
    An introduction to laboratory automation / Víctor Cerdá, Guillermo
Ramis.
        p.   cm. — (Chemical analysis, ISSN 0069-2883 ; v. 110)
    "A Wiley-Interscience publication."
    Includes bibliographical references.
    ISBN 0-471-61818-7
    1. Chemical laboratories—Automation.   I. Ramis, Guillermo.
II. Title.   III. Series.
QD51.C43   1990
542'.8—dc20   90-35305
                                        CIP

Printed in the United States of America

10 9 8 7 6 5 4 3 2 1

QD51
C43
1990
CHEM

PREFACE

The microprocessor revolution and, particularly, the introduction of personal computers are opening new exciting possibilities for the professionals of experimental sciences. Our experience is that some basic concepts of analog and digital electronics, microprocessors, data acquisition interfaces, digital communications, and robotics allow us to explore the limits of all kinds of laboratory equipment. A high level of automation in all types of research and routine tasks can easily be implemented. Furthermore, most of these possibilities are within the economic means of most laboratories, since they require very little additional investment.

However, automation means that the scientist must be trained in these and related interdisciplinary fields, which, in most cases, are different from their own fields and were not included in the curriculum of their professional studies. Moreover, technical books on these subjects are frequently hard to understand by newcomers to the field because of the specialized nomenclature and because the books usually take for granted that the reader has experience in the field.

Our aim in this book is to provide all the necessary knowledge to implement data acquisition and control systems for laboratory automation. Thus, some basic concepts of analog and digital electronics will help the scientist to select components with appropriate quality to avoid the introduction of parasitic signals and excessive background noise. An elementary training in electronics is enough for constructing simple circuits, which frequently are difficult to find on the market with the required specifications. A few basic electronic concepts are also needed to design systems able to be controlled by a computer, as well as to make proper use of the commercially available articulated arms.

Mastery of a high-level scientific computer language, such as FORTRAN or an advanced BASIC, is also necessary to develop the software that must support the mechanical operations and data acquisition and treatment. In addition, some knowledge of the principles of microprocessor architecture is useful to understand the programming, to take advantage of the use of memory maps (thus avoiding overlapping of resident programs), and to make proper use of microprocessor ports. Finally, when the objective is to intercommunicate automated instruments and to fully automate a laboratory procedure, the most popular communication protocols should be known.

This book is written from a practical point of view, and therefore it contains many examples with technical details, circuits, and programs. Most of them have been used in our own laboratory at the University of the Balearic Islands, and some of them are used on a regular basis for practical purposes, such as control of water quality and atmospheric pollution, at the attached Laboratory of Analysis and Assays.

This book was made possible thanks to the help of many people working in the Laboratory Automation Research Group at the University of the Balearic Islands. Over the years, they developed most of the software described here, implemented the hardware, and developed the applications. We are very grateful to all of them and will always be indebted to them.

Additional thanks are due to the manufacturers of equipment and software who gave their permission to reproduce data from their publications. We are particularly grateful to CRISON Instruments, which provided us with valuable information.

We are also indebted to Professor M. Celia García Alvarez-Coque, who revised the manuscript and made many valuable suggestions. She also, with infinite patience, did most of the drawing and typing work. Without her help and continuous support it would not have been possible to deliver the manuscript for this book to the publisher on time.

Finally, the authors thank DGICYT of Spain for financial support of Projects PA86-0033 and PB87-999-02-2, both related to the subjects treated here.

We hope that this book will fully acquaint readers with the field of automation sciences allowing them to bypass the self-educational process that we went through.

VÍCTOR CERDÁ
GUILLERMO RAMIS

Palma de Mallorca, Spain
Valencia, Spain
July 1990

CONTENTS

CHAPTER

1

AUTOMATION AND CHEMISTRY

1.1. INTRODUCTION

Modern automatic devices and automation systems are based on microelectronic components. The huge development of microelectronics and automation systems is changing the way we work, thus freeing us from tedious, iterative tasks and allowing us to devote our efforts to creative activities. Therefore, most of our time can be used in the study of more efficient working strategies and in research planning.

Furthermore, automation permits us to carry out new types of experiments, such as those requiring acquisition of large amounts of data in a short time, simultaneous control of a large number of parameters, or long time periods for completion.

Before introduction of the microprocessor, automatic systems were built up with mechanical devices. Such automatic systems were not flexible. They were designed to accomplish a specific task and usually could not be reused for other purposes. This situation changed dramatically when the microprocessor appeared. Today, most automatic systems are controlled by microprocessors. These complex and inexpensive devices make possible the acquisition, storage, and treatment of an enormous amount of information. The interest in handling large volumes of data increases as the development and popularization of analytical instrumentation make them easier to obtain. Large amounts of data can be generated in a short time by instruments capable of performing multicomponent analysis, such as chromatographs or multichannel spectrometers. Finally, important conclusions can be drawn when multivariant statistical methods are applied to the data.

On the other hand, microprocessors provide many powerful and flexible ways of achieving full laboratory automation. However, instead of building up automatic systems with raw microprocessors, a microprocessor-based system, such as a computer, may be used. Today, automation is largely facilitated by the use of personal computers, which are increasingly more powerful, compact, and inexpensive. Also, they can communicate with

1

scientific instruments and all kinds of devices using standard interfaces and may be programmed using high-level languages. Many laboratory instruments are provided with the appropriate interfaces for communication with microcomputers. This permits one to implement many automatically controlled experiments by reusing and rearranging the same elements.

1.2. AUTOMATION OF THE DIFFERENT STEPS OF AN ANALYTICAL CONTROL PROCESS

To analyze a sample is a control process. Depending on the objectives to be reached and the difficulties involved, automation of this control process can be simple or very complex. Analytical methods are usually classified by the type of physical measurement performed. However, to take a measurement is only one of the many steps found in an analytical process and frequently it is easy to automate. More sophisticated automatic devices are necessary to perform the complex mechanical movements required for operations like sampling, weighing, solving, or dosifying. Some of these processes are still inefficiently automated or have not been automated at all.

The steps to follow in an analytical control process, as well as some of the approaches currently used to automate these steps, are indicated in Table 1.1. In some cases the solution to an automation problem can be simple, for example, a dosifier valve moved by a motor. Other problems require the use of sophisticated elements such as articulated arms. A specialized articulated arm or robot needs a logical support (software) and several complementary mechanical and electrical devices.

A computer can be used to control and harmonize a wide variety of tasks. Sometimes "dedicated" microprocessors are used between the instruments and the central computer, which is also called the "host." Thus, for instance, it is usual to control the movement of the components of the measuring instrument and the acquisition and transfer of data to the host with dedicated microprocessors. On the other hand, the treatment and evaluation of the data, as well as the task of taking decisions about the process, can be done by the host computer. Thus, a continuous feedback between host and process under control can be established.

An example of this is given in Figure 1.1. Production volume and product quality in a distillation plant are regulated in accordance with the customers' pending orders. The central computer controls the sampling frequency at the head of the distillation column by means of valve V_1. The components are separated and determined by gas chromatography, the

Table 1.1. Automation of the Steps of an Analytical Control Process

Operation	Remarks	Equipment
Sampling	Solids	Robot
	Liquids	Dosifier
	Gases	Gas trap
Physical	Size reduction	Robot, mill, dosifier
sample	Drying	Robot, oven, IR lamp
conditioning	Solving	Robot, vortex, heater
	Filtering	Robot, filter
	Centrifuging	Robot, centrifuge
Chemical	Weighing	Robot, balance
sample	Aliquot taking	Robot, dosifier, pipette
conditioning	Injection	Robot, injector
	Solvent extraction	Robot, extraction system
	pH adjustment	Robot, mixer, reactor
	Masking	Robot, mixer, reactor
	Derivatization	Robot, mixer, reactor
Calibration	Calibration	Robot, rack with standards
and	Measurement	Robot, sensor
measuring	Transduction	Transducer
	Amplification	Amplifier
	Signal conditioning	Noise filter, digitalization
	Data storage	Dedicated microprocessor and host computer
Data treatment, evaluation of results, and decisionmaking		

data are digitalized in an analog-to-digital converter and finally enter the computer through an RS232C protocol. The computer integrates peak areas, recognizes interferences, and quantitates the components of interest.

System calibration is done by controlling valve V_2, which selects between sample and standards. In accordance with the results, the computer regulates the position of the electrovalve V_3, by using one of the analog or digital ports of the interface. Column backflow is thus modified to adjust the product quality to our requirements. Production volume can be regulated through feeding valve V_4. The amount of energy supplied, as well as the purge flow, must also be regulated according to the feeding flow. This is done by means of valves V_5 and V_6, respectively.

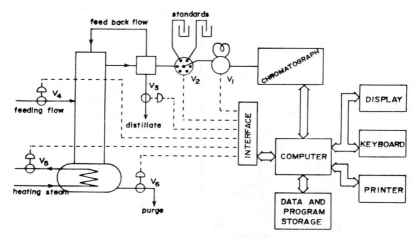

Figure 1.1. Automation of a distillation plant.

Continuous monitoring of the system can be implemented through a cathode ray tube (CRT) or TV monitor or by using a printer or a paper chart recorder. Data can also be stored on a magnetic support, even on the same one containing the control software. New ordering information or changes in control programming are done through a console operation.

This control process contains all the steps indicated in Table 1.1, that is, sampling, sample preparation (e.g., filtering and derivatization), aliquot taking, calibration and measuring, data treatment and evaluation, and decisionmaking.

Automation of such a variety of operations requires wide interdisciplinary knowledge, including analog and digital electronics, microprocessor and computer architecture, digital communications, and robotics.

1.3. COMPUTER OPERATION MODES

Modern automation systems are almost exclusively based on computers. The three types of relationship among experiments, computers, and humans which are usually distinguished (i.e. off-line, on-line, and in-line operation) are depictured in Figure 1.2.

Off-line operation of a computer is a mode in which the computer is utilized almost exclusively as a computational element. Data are collected from the experiment by a manual process, without the aid of the computer. Later, the data must be transferred manually to the computer for calcula-

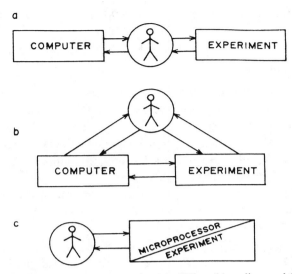

Figure 1.2. Computer operation modes: (a) off-line, (b) on-line, and (c) in-line.

tions. The results are retrieved, studied, and used for the design of new experiments.

In on-line operation, a link exists between experiments and computer. Data move directly from the experiment to the computer, where they are processed and decisions are made. These decisions are based on the program, on the information given to the computer at the time of the experiment, and on the processed data. Resulting commands or data can be returned to the experiment to control it.

The importance of the on-line processes has increased recently, particularly since personal computers (PCs) have become inexpensive and popular. The use of PCs for on-line operation permits the control of very short lifetime operations (milliseconds and less than that) on a real-time basis. This book is concerned mostly with on-line control of experiments of this type.

Instrumentation manufacturers soon became aware of the advantages that the incorporation of computers to other instruments could have and, consequently, in-line operation instruments appeared. In the in-line operation mode, computer and instrument are fully integrated. However, in-line operation has some drawbacks. Thus, for instance, the manufacturers do not reveal the exact nature of the integration, curve fitting, or other calculations, particularly when corrections for interferences or nonlinearities are involved, and the programs cannot be modified by the users.

1.4. SETTING OUT AN AUTOMATION PROBLEM

To set out an automation problem, the following aspects should be considered:

1. Definition of the problem and objectives.
2. Analytical methods available (physical, chemical, or physico-chemical).
3. Control of mechanical systems through interfaces with the computer.
4. Communication requirements between the elements and type of protocol to be used (RS232C, IEEE488).
5. Sensoring or transducting of physical signals to electrical, and transformation of analog signals to a digital form (via A/D converters) or vice versa (D/A converters).

A great variety of the required elements cited above are currently found on the market at very low prices. Standard cards designed to carry out several functions with interesting performances are widely available. Some of them, such as the RS232C interfaces, are usual components of standard PC configurations.

Today, preconditioning for automation is a relevant feature of commercial instrumentation. When a new piece of equipment must be purchased one must consider whether an adequate I/O interface for communication with other elements is available. This will make the instrument or the accessory capable of sending data and of receiving commands. Unfortunately, many commercial instruments have a closed design. Furthermore, some manufacturers still insist on putting obstacles to compatibility by using singular diskette formatting and unusual communication protocols. Of course, information about them is never given to the customer, who is bound to buy new accessories from the same manufacturer. However, since many costumers are becoming better informed and scientists and technicians are being introduced to the theory and practice of automation, this situation is changing. Properly advised buyers look for modular and versatile pieces of equipment which can easily be intercommunicated and thus reused to perform many different tasks.

BASIC CONCEPTS AND ELEMENTS OF ANALOG ELECTRONICS

2.1. INTRODUCTION

In this chapter, it is assumed that the reader has an elementary knowledge of electronics; otherwise the following pages can be skipped. Some important concepts on linear electronics and basic microelectronic devices are described. Any textbook on electronics is appropriate for further studies. Although some basic concepts on digital electronics are also introduced in this chapter, the subject is developed in Chapter 3.

Voltage and current are basic magnitudes in any electric circuit. Voltage—the difference of potential or electromotive force, V—is the work done to move a unit of positive charge from a lower to a higher potential. The unit of measurement for voltage is the volt (symbol V). Usual voltages in analytical instruments range from mV (as in potentiometry) to kV (as in x-ray lamps). When large ranges of voltage (or other magnitudes) are involved, a given value is taken as a reference, V_r, and the voltages of interest are given in decibels as follows:

$$dB = 20 \log(V/V_r). \qquad (2.1)$$

The current, I, is the rate of flow of electric charge at a point. Currents are measured in amperes (A), it being usual to find currents ranging from nA (as in polarography) to amperes (as in electrogravimetry). The power (P) consumed by a device is measured in watts (symbol W) and is given by $P = VI$.

A voltage source is a device that maintains a fixed voltage drop across its output terminals, regardless of load resistance. Analogously, a current source must maintain a constant current through an external circuit, regardless of load resistance or applied voltage. Voltage and current sources are characterized by the output range they can supply, as well as by the maximum load that can safely be supported.

A usual task in the design of electronic circuits is to relate currents and voltages along the components. Thus, in a resistor, the current is proportional to V; in a capacitor, it is proportional to the rate of change of V; in a diode, it flows only in one direction; a thermistor behaves as a resistor but its properties change with temperature, and so on. The current flowing through a material is a function of the voltage drop across it. This function can be linear, as in resistors, or nonlinear, as in diodes and transistors.

2.2. SIMPLIFIED REPRESENTATION OF CIRCUITS

Complex electronic circuits can easily be described by using standard symbols that represent electronic components. Some elementary symbols are shown in Figure 2.1. Other symbols are given later, together with the description of each component.

One of the terminals of a circuit is frequently the ground; that is, it is connected to the soil and its potential is therefore zero. The schemes of the circuits are usually simplified by replacing a whole line by a few ground symbols. The ground terminal may also not really be connected to the soil. In this case, it is said that the ground is floating.

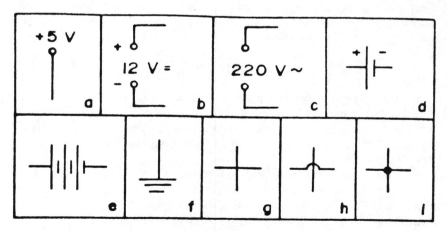

Figure 2.1. Some elementary symbols: (a) voltage supply; (b) dc power supply; (c) ac power supply; (d) battery (the signs are usually omitted); (e) multiple battery; (f) connection to ground; (g) and (h) not connected conductors; (i) connected conductors.

2.3. RESISTORS

2.3.1. Introduction

A resistor, made up of some kind of conducting material (e.g., graphite, metal films, or wires), is characterized by its resistance, R, which is the slope of the plot of V versus I:

$$R = \frac{V}{I}. \tag{2.2}$$

This relationship is known as Ohm's law. Resistances are measured in ohms (Ω). Typically, resistors have values ranging from 1 Ω to several megohms ($M\Omega$). It is usual in electronics to simplify the way resistances are expressed by writing, for instance, 1K instead of 1 $k\Omega$, or 1K5 instead of 1.5 $k\Omega$.

Resistors are also characterized by the amount of power they can safely dissipate (e.g., 0.25 W, 1 W), by their tolerance with respect to the nominal resistance, their temperature coefficient, or stability with time.

Regulable resistors are usually called potentiometers or "pots." The symbols for fixed and regulable resistors are shown in Figure 2.2. Potentiometers are available in several forms, such as with a sliding or a rotating wiper, and with a high precision (e.g., 10-turn rotating wiper), with or without a blocking mechanism. Trimmer potentiometers are regulated with a screw, thus avoiding accidental movements of the wiper.

2.3.2. Voltage Divider

Resistors are used to convert a voltage to a current and vice versa, as well as to reduce a voltage to the value required to feed a component or a part of a circuit. The latter operation is done with a common and useful combina-

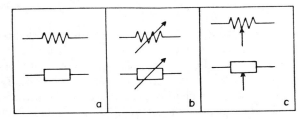

Figure 2.2. Resistor symbols: (a) fixed; (b) regulable; (c) regulable with both wiper and ends connected to a circuit.

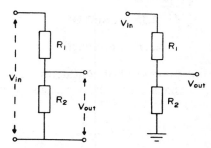

Figure 2.3. Voltage divider.

tion of resistors called a voltage divider (see Figure 2.3). The voltage drop across the input terminals, V_{in}, is reduced by the resistors R_1 and R_2 to the value V_{out}. The relationship between input and output is given by

$$V_{out} = \frac{R_2}{R_1 + R_2} \, V_{in}.$$ (2.3)

Both parts of Figure 2.3 correspond to identical circuits, with the difference that in the circuit of the right, one of the voltages is grounded and therefore its value is zero.

2.4. SIGNALS AND NOISE

The most common electronic signals are sinusoidal. The great advantage of using sinusoidal signals is derived from the fact that they are the solution of certain linear differential equations that happen to describe many phenomena in nature, including the properties of the linear circuits. A circuit is linear if the output, when driven by the sum of two output signals, equals the sum of its individual outputs when driven by each input signal in turn.

Several types of signal commonly used in electronics and electrochemistry are ramps, sawtooths, triangle waves, square waves, and pulse waves. These signals are schematized in Figure 2.4. A ramp is a voltage rising or falling at a constant rate. A periodic ramp, which rises or falls in a given time and goes back suddenly to the initial value, is called a sawtooth. When the signal gets back to the initial value following a reversed ramp, a triangle wave is obtained. A square wave is a signal that periodically goes back and forth between two different constant values with the same lifetimes. When the lifetimes of the two levels are different, a periodic pulse or pulse train is

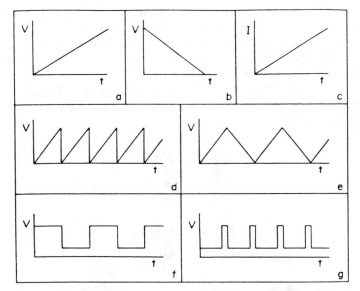

Figure 2.4. Common types of signal: (a) voltage ramp; (b) reversed voltage ramp; (c) current ramp; (d) voltage sawtooth; (e) triangle wave; (f) square wave; (g) periodic pulse.

achieved. Periodic signals are characterized by the amplitude, or peak-to-peak oscillation, and by the frequency. Typical amplitudes and frequencies range from μV to V and from Hz to MHz, respectively.

Signals of interest are always mixed with noise. Noise can be characterized by its frequency and amplitude distribution, which constitute the spectrum of the noise. One of the most common kinds of noise is white gaussian noise. A noise is said to be white when it carries the same amount of power at any frequency, within some band of frequencies. The noise is gaussian when a large number of instantaneous measurements of its amplitude give a gaussian distribution. For instance, the noise generated by a resistor is white gaussian.

2.5. LOGIC LEVELS

Digital electronics is based on Boolean logic: two voltage levels, simply called low and high, correspond to the 0 or false, and to the 1 or true logic states, respectively. A negative logic, where the logic true is 0 and the false is 1, is also used. The voltage levels used in three common families of digital logics—TTL, RS, and CMOS—are shown in Figure 2.5. The shaded areas show the specified ranges of output voltages that a logic high or low is guaranteed to fall within. Accurate voltages are not necessary in digital electronics, since only two different possible states have to be distinguished.

Pulse and square waves are extensively used in digital electronics, the two voltage levels representing one of the two possible logic states at a given point of the circuit.

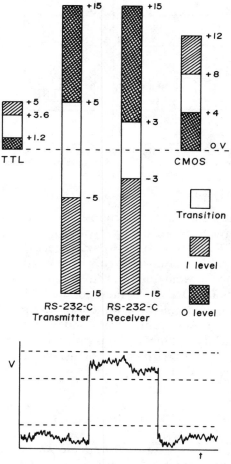

Figure 2.5. Popular families of digital logics. Lower part: A TTL level 0 changes to 1 and then back to 0 without being affected by noise and voltage drift.

2.6. CAPACITORS, RC CIRCUITS, AND APPLICATIONS

2.6.1. Capacitors

Capacitors are essential components of most circuits. They are used to generate waveforms and in filtering, blocking, and bypassing tasks. They are also used to build up integrating and differentiating circuits.

To a first approximation, capacitors might be considered as frequency-dependent resistors. In a capacitor, the accumulated charge, Q, is proportional to the voltage across its terminals, V:

$$Q = CV. \tag{2.4}$$

Taking the derivative, we obtain

$$I = C \frac{dV}{dt}. \tag{2.5}$$

The current is therefore proportional to the rate of change of the voltage. The proportionality constant, C, called the capacity, is given in farads. Capacities range typically from picofarads (pF) to microfarads (μF). These are the usual units, nF seldom being used. Instead, the unit kpF, frequently written with a simple K, is used. The different types of capacitors and main characteristics are summarized in Table 2.1.

Table 2.1. Types of Capacitors

Type	Capacitance Range		Remarks
Mica	1 pF	– 0.01 µF	Excellent
Tubular ceramic	0.5	– 100 pF	—
Ceramic	10 pF	– 1 µF	Small, poor temperature stability, inexpensive
Mylar	1 K	– 10 µF	Good, poor temperature stability, inexpensive
Polystyrene	10 pF	– 0.1 µF	Large, high quality
Polycarbonate	100 pF	– 10 µF	High quality
Glass	10 pF	– 1 K	Long-term stability

Table 2.1. (*Continued*)

Type	Capacitance Range	Remarks
Porcelain	100 pF – 0.1 µF	Good, inexpensive, long-term stability
Tantalum	0.1 – 500 µF	Poor accuracy and poor temperature stability, polarized, small
Electrolytic	0.1 µF – 0.2 F	Bad accuracy and bad temperature stability, high leakage current, polarized, short life; not recommended except in power supply filters
Oil	0.1 – 20 µF	Large, long life, good for high voltages

2.6.2.   RC Circuits

The combination of a resistor and a capacitor is called an RC circuit (see Figure 2.6a). When dealing with RC circuits, the rate of charge and discharge of the capacitor through the resistor is of great importance. In the RC circuit of Figure 2.6b the charge accumulated by the capacitor is discharged through the resistor when the switch is turned on. From Ohm's rule and Equation (2.5), we have

$$C \frac{dV}{dt} = I = - \frac{V}{R}. \tag{2.6}$$

This differential equation has the solution

$$V = Ae^{-t/RC}, \tag{2.7}$$

which is represented in Figure 2.6d. The product RC is called the time constant of the circuit. For R in ohms and C in farads, the RC product is given in seconds.

Similarly, for the circuit of Figure 2.6c, the charge equation is

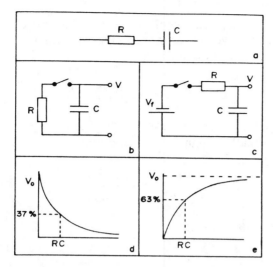

Figure 2.6. (a) Scheme of a RC circuit; (b) and (d) the circuit in discharge; (c) and (e) the circuit in charge.

$$I = C \frac{dV}{dt} = \frac{V_f - V}{R}, \tag{2.8}$$

which upon integration gives

$$V = V_f + A e^{-t/RC}. \tag{2.9}$$

As can easily be calculated, for a tolerance of $\approx 1\%$, V reaches V_f in 5 time constants ($5RC$).

2.6.3. Differentiators and Integrators

RC circuits are used to build up differentiating and integrating circuits. Differentiators are used in analytical instrumentation for detecting leading

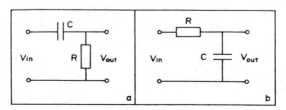

Figure 2.7. (a) RC differentiator and (b) RC integrator.

and tailing edges of pulse signals. A differentiating circuit is shown in Figure 2.7a. Since the voltage drop across the capacitor is $V_{in} - V_{out}$, the discharge equation of the circuit is

$$I = C \frac{d(V_{in} - V_{out})}{dt} = \frac{V_{out}}{R}. \tag{2.10}$$

If R and C are small enough for the condition

$$\frac{dV_{out}}{dt} << \frac{dV_{in}}{dt} \tag{2.11}$$

to be acceptable, then it will be

$$C \frac{dV_{in}}{dt} = \frac{V_{out}}{R} \tag{2.12}$$

and

$$V_{out}(t) = RC \frac{dV_{in}(t)}{dt}, \tag{2.13}$$

which means that the output is proportional to the derivative of the input signal.

Integrators are also very useful in analog computation and, in general, in analytical instrumentation. An integrating circuit is shown in Figure 2.7b. Since the voltage drop across R is $V_{in} - V_{out}$, the discharge equation is

$$I = C \frac{dV_{out}}{dt} = \frac{V_{in} - V_{out}}{R}. \tag{2.14}$$

Now if RC is large enough, $V_{out} << V_{in}$, and

$$C \frac{dV_{out}}{dt} = \frac{V_{in}}{R}. \tag{2.15}$$

Finally,

$$V_{out}(t) = \frac{1}{RC} \int_0^t V_{in}(t)dt + K, \tag{2.16}$$

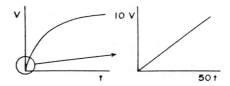

Figure 2.8. Output of a RC integrator as an approximately linear ramp generator.

which means that the circuit integrates the input signal over a given time. As is shown later in this chapter, integrators can also be constructed using operational amplifiers, with the advantage that the restriction $V_{out} << V_{in}$ is not required.

2.6.4.   Ramp Generators

Another important application of RC circuits is ramp generation. Timing circuits, waveform generators, analog-to-digital converters, and many other devices contain ramp generators. A ramp generator is built up with a resistor and a capacitor mounted as in an integrator; however, only the very first part of the integrated curve is used. As shown in Figure 2.8, the beginning of the integral of a constant is almost a linear ramp. The use of a very large RC value makes it possible to pick up an approximately linear part of the exponential curve.

A ramp can also be generated by feeding the circuit of Figure 2.7b with a source of constant current. An approximately constant current can be obtained by applying a large voltage to a large resistor. From Equation (2.5), we have

$$\frac{dV}{dt} = \frac{I}{C},\qquad (2.17)$$

which means that by using a constant current source, a perfectly linear ramp is obtained. This is a simple and frequent approach in ramp generation. The outputs of both approaches are compared in Figure 2.9.

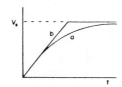

Figure 2.9. Comparison of the outputs of a ramp generator fed with a constant voltage (curve a) and with a constant current source (curve b).

2.7.   DIODES, LEDs, AND ZENERS

Resistors and capacitors are passive linear elements. A passive element does not contain any internal source of power, and in a linear element doubling of the applied voltage produces doubling of the current. Diodes are also passive, but they are not linear.

The typical response curve of a *diode* is shown in Figure 2.10. In order to realize another important feature of diodes, attention should be paid to the change of scale in this figure. As can be deduced, a diode is a good approximation to an ideal one-way conductor. The symbol for a diode contains an arrow pointing in the direction of forward current flow. Frequently, the forward voltage drops between 0.5 and 0.8 V, which must be taken into account in some applications.

There are some special types of diode, such as the LEDs and the zeners. A *light-emitting diode,* or *LED*, is a solid-state lamp. It is a *pn*-junction diode that is specially designed to produce a light output when energized. Most common LEDs used in display applications require 15–20 mA. LEDs are available in a number of light color emissions and shapes. The seven-segment numerical display is a very common form of numerical output that is frequently constructed with seven bar-shaped LEDs. By lighting selected segments, any of the hexadecimal digits 0–9, A, B, C, D, E, and F can be formed. By connecting the bits of a computer output, the digits can be turned on as needed.

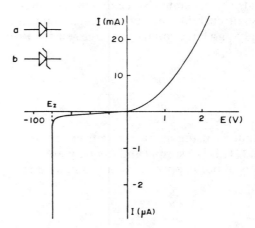

Figure 2.10. Symbols of a diode (a) and a zener (b). Response of both devices. Attention to the scale changes must be paid. E_z = zener potential.

A *zener* is a diode designed to work under reversed polarization, that is, in the so-called zener region. As shown in Figure 2.10, when a given voltage is exceeded, a zener suddenly permits a current to flow, thus limiting the applied voltage. Zeners are therefore used as voltage standards and constitute by themselves simple voltage regulators. A great variety of them are available.

A zener working as a voltage regulator is shown in Figure 2.11. The current flowing through the resistor R must be higher than the load current for a current to flow to the ground through the zener:

$$I_{in} = \frac{V_{in} - V_{out}}{R} > I_{out}. \tag{2.18}$$

Otherwise, the zener will not operate.

The zener maintains a constant output ($V_{out} = E_z$) at the cost of the power being dissipated through the zener, which is given by

$$P_z = I_z E_z = \left(\frac{V_{in} - V_{out}}{R} - I_{out}\right) E_z. \tag{2.19}$$

For safe zener operation, there is a maximum power that should not be exceeded and that must be reached when I_{out} equals zero. Consequently, this simple regulator must be designed for a maximum load current. When a higher current is demanded, the power dissipated at the zener drops to zero and the voltage fails to be regulated.

The use of a zener as a voltage regulator has still other drawbacks, such as the impossibility of adjusting V_{out}, which is given by the selected zener; the ripple coming from the ac source is not eliminated and, finally, too much power is lost when the load is small. Better voltage regulators, some of them based on combinations of zeners with other elements, are explained later in this chapter.

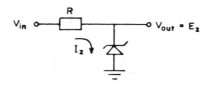

Figure 2.11. A zener working as a voltage regulator.

2.8. RECTIFICATION AND FILTERING

One of the simplest and most important applications of diodes is current rectification. A rectifier is a circuit that changes an ac to a dc. Thus, for instance, a half-wave rectifier can be built up with a single diode, as shown in Figure 2.12. A full-wave rectifier built up with a typical four-diode bridge is shown in Figure 2.13. The small gaps at zero voltage are due to the forward voltage drop of the diodes. Diode bridges are available as integrated circuits (ICs, see Section 2.9)—four diodes packed in a small and inexpensive chip.

Rectifiers built up with only diodes remove one of the polarities of an ac supply but, since the input is sinuisoidal, the output always has a "ripple" that must be smoothed to obtain a better, approximately constant, signal. This is done with a low-pass filter. The rectifier circuit of Figure 2.14 includes a simple filter, built up with a resistor and a capacitor. In order to ensure a small ripple, the capacitor value must fulfill the condition

$$R_l C \gg \frac{1}{f},$$

where R_l is the load and f is the ripple frequency—twice the input ac frequency (e.g., 2×60 Hz in America and 2×50 Hz in Europe).

The circuit in Figure 2.15 is called a center-tapped full-wave rectifier. The output voltage is half the value obtained from a bridge rectifier connected to the ends of the secondary of the transformer.

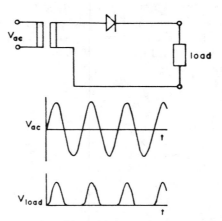

Figure 2.12. Scheme (upper part), and input and output (lower parts) of a single diode rectifier. The rectangles at the left represent the induction coils of the transformer.

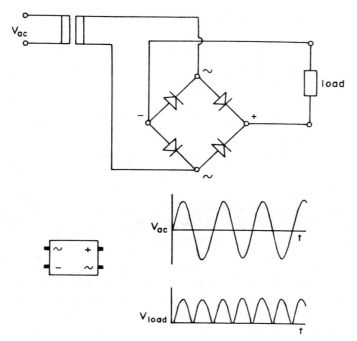

Figure 2.13. Scheme, chip pin connections, and input and output of a bridge rectifier.

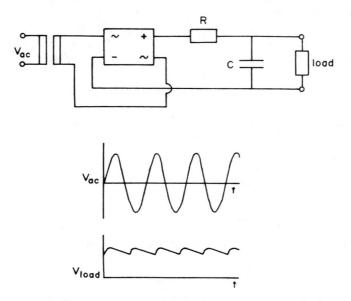

Figure 2.14. Bridge rectifier with low-pass filter, input and output.

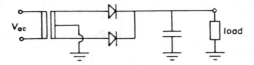

Figure 2.15. A center tapped full-wave rectifier.

2.9. INTEGRATED CIRCUITS

An integrated circuit or IC is a combination of electronic components packed into a single support and designed to perform an electronic task or function. Usually, the support is an inexpensive silicon chip. Today, the techniques of very large integration scale (VLIS) permit hundreds of thousands of miniaturized components to be packed in a small 25 mm^2 chip.

Simple ICs, used as voltage regulators or operational amplifiers, have the shape and appearance shown in Figure 2.16 (left), with three or four pins for external connections. More complex ICs are built up into standard 14-, 16- or 40-pin chips (see Figure 2.16, central part), which are designed to be mounted on 14-, 16-, or 40-connection rectangular bases. In most cases, only a few pins are connected to internal components, the rest of them being nonconnected (NC). A small dip or point indicates where pin number 1 is located. The rest of the pins are numbered counterclockwise. In electronic schemes, the indication of pin number 1, as well as the pins not used are omitted. The other pins are properly numbered, but usually they are not shown in the same order as they appear in the real chip (see Figure 2.16, right). In this way, a considerable simplification of the schemes is achieved.

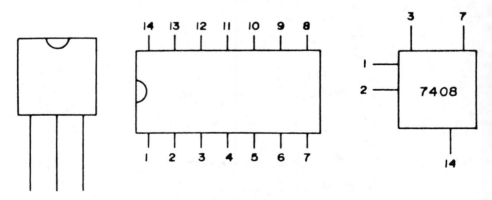

Figure 2.16. A 3-pin and a 14-pin chip. Schematic representation of the 7408 IC with only 5 pins connected to the external circuitry.

2.10. REGULATORS

2.10.1. Introduction

The ripple amplitude of a dc current, obtained by rectification of an ac, can be reduced to any desired level by using sufficiently large capacitors. However, this approach has several disadvantages: prohibitively bulky and expensive capacitors may be needed; input line fluctuations are transmitted to the output voltage and, due to the finite internal resistances of the elements of the rectifying and filtering circuits (transformer, diodes, resistor, etc.), the output voltage changes with the load. It is better to use a regulator together with a capacitor as explained next.

Regulators are active feedback circuits that give a constant dc voltage. Voltage regulators, available as inexpensive chips (Figure 2.17), are used almost universally as power supplies in electronic circuits. A power supply built up with a voltage regulator has excellent properties as a voltage source. Voltage regulators can easily be adjusted to give the desired voltage and are internally self-protected against short circuits and overheating.

A regulator is mounted as shown in Figure 2.18. The load, connected to the output and ground terminals of the regulator, is kept at a constant voltage, at the cost of some power, which is dissipated as heat by the regulator through the input–ground path. The capacitor across the output serves to further diminish the residual ripple.

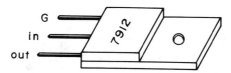

Figure 2.17. A voltage regulator showing the attached metallic plate for heat dissipation. The hole is optionally used to screw the plate to a larger piece of metal.

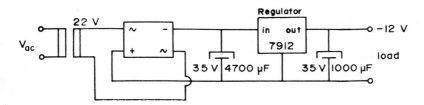

Figure 2.18. Voltage source built up with a voltage regulator.

2.10.2.   Three-Terminal Regulators

A popular family of regulators are the 78xx and 79xx series. The last two digits make reference to the output voltage and may be 05, 06, 08, 10, 12, 15, 18, or 24, whereas the first two digits indicate the sign of the output, positive in the 78xx series and negative in the 79xx. Maximum input voltage is 35 V. When properly refrigerated, safe load current can be as high as 1 A. The medium and low series, 78Mxx and 78Lxx, can only support currents of 0.5 and 0.1 A, respectively.

A very useful voltage source, built up with three 3-terminal regulators, is shown in Figure 2.19. This source may be used to provide voltages for two families of digital logics, the TTL and the CMOS (see Section 2.5) and also for RS232C voltage adapters (see Chapter 6). It can also be used as a ± 12 V supply for operational amplifiers. The capacitors C_2 across the outputs improve the transient response, smoothing edges and spikes, and keep the impedance low at high frequencies.

2.10.3.   Four-Terminal Regulators

Output voltages can be adjusted to several values if a four-terminal regulator is used. The four terminals are called input, output, ground, and control. These devices give a constant voltage between the output and control terminals, and the desired variable voltage is obtained between the output and ground terminals by using an external voltage divider designed by the user.

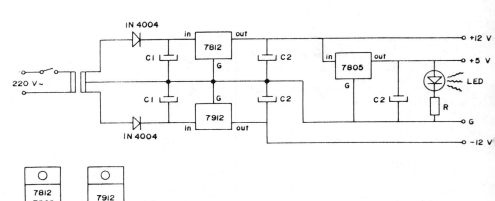

Figure 2.19. A voltage source, built up with three regulators, which supplies the useful voltages +5, +12, and −12 V.

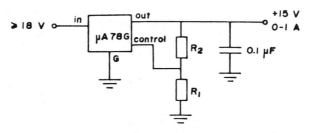

Figure 2.20. Regulable voltage source based on a four-terminal regulator.

An example of such a circuit is given in Figure 2.20. The μA78G regulator maintains the resistor R_2 at 5 V. By proper selection of the values of R_1 and R_2, +15 V are regulated between output and ground. Calculations are done by simply rearranging the voltage divider law [Equation (2.3)]:

$$V_{out} = V_{R_2} \frac{R_1 + R_2}{R_2}. \tag{2.20}$$

A trimmer potentiometer can be used for R_1, to precisely set the desired voltage. Other circuits that supply regulable voltages and that are based on operational amplifiers are described later in this chapter.

2.10.4. Voltage Regulation of Large Load Currents

Currents higher than those supported by regulators can be handled by using an auxiliary transistor (see Section 2.11). A circuit is shown in Figure 2.21. For load currents smaller than 100 mA, the circuit works as before, but for a larger current, the voltage drop across R_1 turns the transistor on. Then, most of the current bypasses the regulator, which supports a current always limited to about 100 mA. As usual, the regulator maintains the output at the right voltage by reducing the input current, which in turn drives the transis-

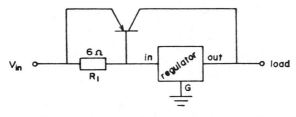

Figure 2.21. An auxiliary transistor permits use of a voltage regulator with a large load current.

tor. Obviously, input voltage must be higher than the output, at least by the drop across the regulator (about 2 V) plus the voltage base–emitter drop, V_{BE} (about 0.6 V).

2.11. TRANSISTORS

2.11.1. Introduction

Transistors are active components of great importance in electronics. Most circuits, such as amplifiers, oscillators, and switches, are based on transistors, and many ICs contain a large number of transistors. Circuit interfacing is also an electronic task usually performed by transistors.

Two main types of transistor are called *npn* and *pnp*. Their symbols and terminals—base (B), collector (C), and emitter (E)—are represented in Figure 2.22. The figure also shows how the base–collector and the base–emitter paths behave like diodes.

In a properly installed *npn* transistor, the collector is maintained at a more positive potential than the emitter. Thus, the formal positive current flows as shown in Figure 2.23. A small current, coming from the base, I_B, flows in the emitter direction. On the other hand, the base–collector path is reverse-biased, and the current cannot flow from the collector toward the base. Instead, a current coming from the collector, I_C, is added to I_B, which can be expressed by

$$I_E = I_B + I_C. \tag{2.21}$$

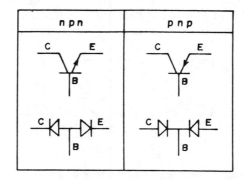

Figure 2.22. Transistors: types, symbols, and function.

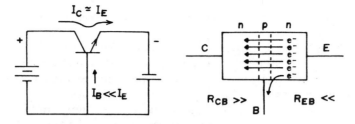

Figure 2.23. An *npn* transistor working as a power amplifier.

Furthermore, I_C is proportional to I_B, which is expressed by

$$I_C = h_{FE}\, I_B,\qquad(2.22)$$

where h_{FE} (also called the β factor) is the current gain. A typical value of h_{FE} can be 100. This means that a small current coming from the base controls a much larger current flowing in the collector–emitter direction: that is, a transistor works as a current amplifier.

Transistors can also work as power amplifiers. Thus, from Figure 2.23, it can be deduced that the resistance through the loop at the left, where the diode junction is found in the reversed direction, is much higher than through the loop of the right; that is,

$$R_C \gg R_E.\qquad(2.23)$$

The power dissipated at each loop is

$$P_C = I_C\, R_C,\qquad(2.24)$$

$$P_E = I_E\, R_E.\qquad(2.25)$$

Since $I_C \approx I_E$,

$$P_C \gg P_E.\qquad(2.26)$$

Current gains for nominally identical individual transistors can be found in a range as large as 50–250. This and the dependence found between current gain and I_C make it unwise to use β as a transistor parameter. Transistors are characterized by other parameters: for example, maximum currents that can safely be supported, breakdown voltages V_{CE} and V_{BE}, maximum power dissipated $I_C V_{CE}$, and safe temperature range.

2.11.2.   Phototransistors and Isolation Techniques

Frequently, different parts of a circuit are required to be isolated from each other. Isolation techniques are frequently used in digital electronics to prevent accidents when a part of a circuit manipulates high voltages, to eliminate noises, or to avoid unexpected electrical leaks.

Optocouplers are very simple isolation devices. As shown in Figure 2.24, an optocoupler contains a LED and a phototransistor (PT) packed into the same chip. PTs are light-driven transistors. They are entirely similar to ordinary transistors but the base input is a photosensitive surface that produces a current when exposed to light. When the LED is activated, light falling on the base leads the transistor to conduction. In the absence of light, the transistor is off. In most cases, the PT is not able to carry enough current, and a second transistor is used. Digital circuits make use of multiple optocouplers for multibit transfer. Their function is to protect computers from noise and catastrophic events.

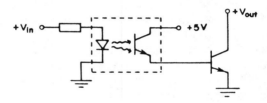

Figure 2.24. Optocoupler and optocoupling circuit.

LEDs and PTs are also used to transmit signals through optical fibers as shown in Figure 2.25, which is particularly useful in noisy environments. LEDs and PTs specially designed to be coupled with optical fibers are available. To pump light through very long transmission paths, He–Ne or diode lasers may be used. Extensive use of these devices is currently found in the telecommunication industry.

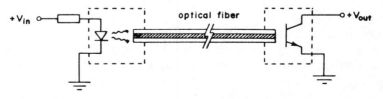

Figure 2.25. Optocoupling through an optical fiber.

2.11.3. Transistor Switching

A simple transistor-based switch is represented in Figure 2.26. When the mechanical switch is open, there is no base current and, as stated by Equation (2.22), there is no collector current either. The lamp is consequently off. When the switch is closed, the base voltage rises to 0.6 V, which is the drop across the diode junction in the direction of forward conduction. The drop across the resistor connected to the base is therefore 9.4 V. Since the resistance is 1K, from Ohm's rule it is deduced that the base current is 9.4 mA. This means that, for a current gain of 100, the current through the collector–emitter path must be as high as 0.94 A and the lamp is on. Transistor switching is useful to switch large currents with a small control current.

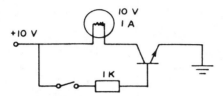

Figure 2.26. A transistor-based switch.

When the load is inductive, such as in a relay, the transistor must be protected with a diode mounted across the load, in the form shown in Figure 2.27. Without the diode, the inductor would swing the collector to a large positive voltage, most likely surpassing the collector–emitter breakdown voltage.

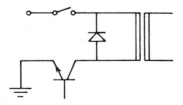

Figure 2.27. Transistor switching of an inductive load.

A useful property of transistor switching is speed. The load circuit is usually switched in much less than a microsecond. Also, by using a series of transistors, many different circuits can be switched at a time with a single control signal.

Thus, for instance, the useful circuit shown in Figure 2.28 permits the switching of a series of relays, using the digital ports of a standard I/O interface of a computer as source of the control signal. When the computer port is set at the high logic level (e.g., at +5 V for a TTL port), current flows through the LED of the optocoupler. This greatly reduces the resistance of the lower arm of the voltage divider, which is given by $R_2 + R_{CE}$, thus polarizing correctly the AC126 transistor base. The AC126 collector–emitter current increases greatly, and the relay is activated by the inductor. Finally, the 220 Ω resistor located in the LED circuit limits the current, thus staying below its breakdown value. The diodes protect the *pnp* transistor against the large induction back-current produced when the relay is turned off. The AC126 transistor can be replaced by a BD140.

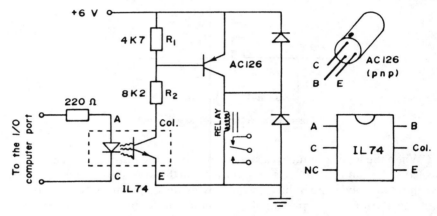

Figure 2.28. Switching circuit showing the pin connections of the AC126 *pnp* transistor and IL74 optocoupler. B is not used.

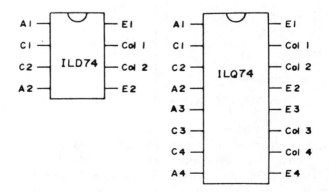

Figure 2.29. Schemes of the ILD74 and ILQ74 optocouplers.

Figure 2.29 shows the schemes of double (ILD74) and quadruple (ILQ74) optocouplers, which are very useful in controlling several relays at a time with a computer.

2.11.4.   Emitter Follower

A transistor mounted as shown in Figure 2.30 is called an emitter follower because the output voltage, which is that of the emitter, follows the input voltage variations, which are those produced at the base. The distance between "pursuer" and "pursued" is constant and equal to the drop across the diode junction in forward direction, 0.6 V, which can be expressed by

$$V_E = V_B - 0.6. \tag{2.27}$$

A voltage regulator, built up with an emitter follower and a zener is represented in Figure 2.31. The zener maintains the base at a constant

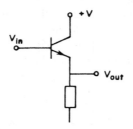

Figure 2.30. Emitter follower.

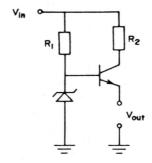

Figure 2.31. A voltage regulator built up with an emitter follower and a zener ($V_{out} = E_Z -$ 0.6 V).

voltage, E_z, and therefore the voltage at the emitter terminal is also constant. In comparison with the use of a single zener, the main advantage is the almost complete independence of the zener current from the load. The transistor base current is small, and thus the dissipation of power at the zener is very small; this is accomplished by adjusting R_1. On the other hand, resistor R_2 is important to protect the transistor from momentary short-circuits at the output.

2.11.5. Field-Effect Transistors (FETs)

The design and properties of field-effect transistors (FETs) are quite different from those of ordinary transistors, which are sometimes called bipolar transistors, to stress the differences. In a FET, a current is controlled by means of an electric field produced by an applied voltage at a control terminal, rather than by a base current. The result is that the control terminal, which is called the gate, virtually draws no current. Also, as a consequence, the input impedance is very high, on the order of 10^{12} or more, which is very useful in many applications. FETs can work as excellent voltage-controlled resistors and as current sources. In addition, many FETs can be packed in small areas, making them excellent for the construction of complex ICs.

The two main types of FET are the junction FETS—or JFETs—and the metal-oxide-semiconductor FETs—or MOSFETs. Each of these is available in two polarities, n-channel and p-channel, corresponding to the *npn*

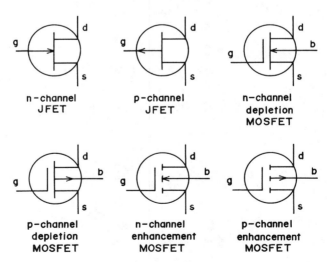

Figure 2.32. Symbols and terminals of FETs; s = source, d = drain, g = gate, b = body or bulk substrate.

and *pnp* bipolar transistors, respectively. Symbols and terminal names are given in Figure 2.32.

The three terminals of a FET—that is, drain, source, and gate—are similar to those of bipolar transistors—collector, emitter, and base, respectively. However, a FET has a single junction between the gate and the drain–source path or "channel." A voltage applied to the gate controls the current that flows through the channel. The n-channel FETs are usually operated with the drain more positive than the source; thus the current flows from drain to source, the reverse being true for p-channel FETs.

However, unlike bipolar transistors, the gate of a JFET must be at ground for the current to flow through the channel. On the other hand, the channel current is cut off when the gate–channel junction is reverse-biased a few volts.

The junction of a JFET is of the same type as in a diode. For this reason, JFETs are always operated with the junction reverse-biased. However, even in this case, some current leaks through the gate. Performance of JFETs is also limited by the possibility of the gate becoming forward-biased with respect to either the source or drain. When this occurs, a drastic reduction of input impedance is produced.

In MOSFETs, gate and channel are truly separated by a thin layer of SiO_2 and current cannot flow through the gate. The channel current is completely controlled by the electric field created by the gate–channel voltage.

The channel ends of a MOSFET are almost symmetrical, thus any of them can be used as a source or as a drain. The "active" terminal, the one that supplies the power, is the drain, and the nonactive terminal (frequently grounded) is the source. With the source at ground, a MOSFET is turned on (brought into conduction) by applying to the gate a voltage of the same sign as the drain.

Due to small manufacturing variations, nominally identical FETs present large differences in their characteristics. Thus, very different gate–channel voltages may be needed to obtain the same channel currents in a pair of FETs. Therefore, when a pair of FETs are used in a differential circuit, a large offset must be applied to equal the responses. However, because of the particular advantages of FETs, manufacturers have made available specially matched pairs of FETs which need small, if any, offset voltages. Matched FETs are very useful in constructing high input impedance front-end stages for comparators, amplifiers, and other circuits.

Frequently, FETs are used as voltage followers. This is a good way to achieve a high input impedance, and a FET follower is commonly used as a first stage in measuring instruments. There are many applications where the signal intrinsically has a high impedance (e.g., glass electrodes) and, in these cases, a FET follower as input stage is a good solution.

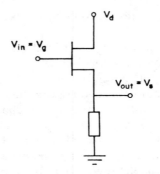

Figure 2.33. A simple FET-based voltage follower. The subscripts d, s, and g are referred to drain, source, and gate, respectively.

A very simple FET-based voltage follower is shown in Figure 2.33. Since the gate current is negligible,

$$V_s = R_{load}\ I_d. \tag{2.28}$$

On the other hand, the gain is given by

$$g = \frac{I_d}{V_g - V_s}. \tag{2.29}$$

Substituting the value of I_d in Equation (2.29), we obtain

$$V_s = \frac{R_{load}\ g}{1 + R_{load}\ g}\ V_g. \tag{2.30}$$

A good follower ($V_s \approx V_g$) is obtained by making $R_{load} \gg 1/g$. The circuit can also be considered as an amplifier with gain approaching unity. The output impedance is given by

$$Z_o = 1/g \tag{2.31}$$

and its value is usually a few hundred ohms for currents of a few mA.

2.12. OPERATIONAL AMPLIFIERS

2.12.1. Introduction

Operational amplifiers (OAs) are high-gain linear devices designed to supply a given transfer function in a circuit. They are built up with a combination of transistors and resistors packed in a chip. OAs are found as essential multifunction components in many circuits. The symbol and pin connections of the popular μA741C OA are represented in Figure 2.34. Some OAs of different types are given in Table 2.2.

On Figure 2.34, the symbols + and − near the input connections do not mean potential signs, but rather a noninverting and inverting behavior. The output moves in a positive direction when the noninverting input (+) becomes more positive than the inverting input (−), and vice versa. In the mini-dip 8-pin chip, pins 4 and 7 serve for powering the OA; for example, with −15 V at pin 4 and +15 V at pin 7. Pins 1 and 5, or offset null terminals,

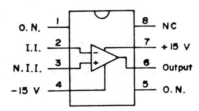

Figure 2.34. The μA741C operational amplifier. O.N. = offset null, I.I. = inverting input, N.I.I. = noninverting input, NC = not connected.

Table 2.2. Some OAs of Different Types

Name	Remarks	Name	Remarks
741S	Fast response	747	Dual 741
MC741N	Low noise	OP-04	Dual, precision
OP-02	Precision	4558	Dual, fast
4132	Low power	TL082	Dual, FET input
LF13741	FET input	MC4741	Quad 741
NE530	Fast response	OP-11	Quad, precision
TL081	FET input	4136	Quad, fast
LM343	High voltage (68 V)	HA4605	Quad, fast
3583	High voltage (300 V)	TL084	Quad, FET input

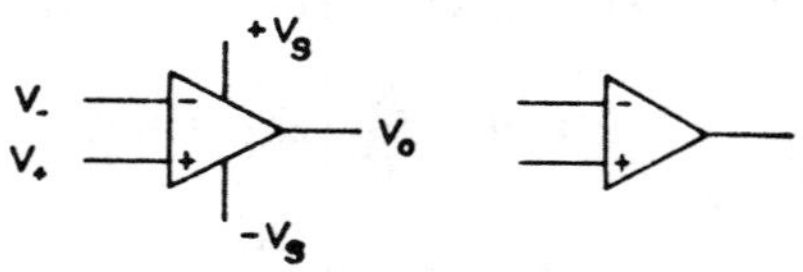

Figure 2.35. Normal and simplified symbol for an OA. V_+ and V_- are the potentials at the noninverting and inverting inputs, respectively; $+V_s$ and $-V_s$ are supply voltages; and V_o is the voltage at the output.

are used to externally correct the small asymmetries that are due to unavoidable imperfections in the manufacture of the OA. The simplified symbol in Figure 2.35, where power supply and offset null connections are not displayed, is frequently used in circuit schemes.

In an OA, several important characteristics, such as frequency response, phase signal change, and gain and transfer functions, are usually established by the amount of output fed back to one of the inputs. Usually, OAs work with negative feedback, which means that the inverting input is partially cancelled by a part of the output. However, in some applications, positive feedback is used.

Negative feedback is implemented with a resistor mounted in a loop coupling terminals 6 and 2 (Figure 2.34). Negative feedback lowers the gain, but, in exchange, it improves other characteristics such as distortion and nonlinearity. As more negative feedback is applied (a smaller resistor in the loop is used), the resultant amplifier characteristics become less dependent on the characteristics of the amplifier working in open-loop conditions (without feedback).

An explanation of the internal circuitry of an OA is beyond the scope of this book. It is more convenient to consider the OA as a black box that behaves in accordance with the following simple rules:

1. The output attempts to do whatever is necessary to make the voltage difference between the inputs zero.
2. Input currents are very small, ranging from μA to pA; thus, for practical purposes, the inputs draw no current.

Amplification is a consequence of rule 1: a small voltage difference between the inputs, which cannot be reduced to zero by the output, gives a much larger output. However, the output cannot exceed the supply voltages. For example, for a μA741C supplied with ±15 V, the output ranges approximately between ±13 V. To keep in mind these rules is important to understanding circuits which include OAs.

The rules given above are only obeyed when the OA is in the active region: that is, inputs and output are not saturated, or, in other words, they must be within specifications for the OA to operate.

2.12.2. The Response of an OA

An ideal OA should supply a linear output, proportional to the voltage difference between the inputs, which is expressed by

$$V_o = (V_+ - V_-)A_d, \tag{2.32}$$

where A_d is the gain, and V_o can be positive or negative, having the same polarity as V_+. A typical, although somewhat idealized, transfer curve of an OA is shown in Figure 2.36. Observe how, when a given value of the difference $V_+ - V_-$ is exceeded, the output almost reaches one of the supply voltages, and Equation (2.32) is not followed. When this occurs, it is said that the output is at positive or negative saturation.

In Figure 2.36, also note that the output does not depend on the absolute values of the inputs, which means that the same output is obtained by applying, for instance, 1.000 and 0.999 V or 5.000 and 4.999 V to the inverting and noninverting inputs, respectively. However, as is shown below, this is not completely true, the output being somewhat higher in the latter case.

The output can be expressed as a linear combination of the inputs:

$$V_o = A_+V_+ + A_-V_-, \tag{2.33}$$

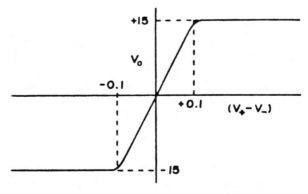

Figure 2.36. Transfer plot of an OA. Voltage supply is ± 15 V and inherent gain is $A = 150$.

where A_+ is the output voltage gain with respect to the noninverting input voltage, V_+, when the inverting input is at ground, the reverse being true for A_-. On the other hand, we can write

$$V_+ = V_c + \tfrac{1}{2} V_d, \tag{2.34}$$

$$V_- = V_c - \tfrac{1}{2} V_d, \tag{2.35}$$

where V_c is the common mode signal—half the sum of V_+ and V_-—and V_d is their difference. By substituting Equations (2.34) and (2.35) in (2.33), we have

$$V_o = A_+(V_c + \tfrac{1}{2} V_d) + A_-(V_c - \tfrac{1}{2} V_d) \tag{2.36}$$

Rearranging, we have

$$V_o = V_c(A_+ + A_-) + \tfrac{1}{2} V_d(A_+ - A_-) = V_c A_c + V_d A_d, \tag{2.37}$$

where A_d is the voltage gain of the differential signal and A_c is the voltage gain of the common mode signal.

A direct measurement of A_d can be obtained by applying the voltages $V_+ = 0.5$ V and $V_- = -0.5$ V to the inputs. Then, $V_d = 1$ V and $V_c = 0$ V, and the output voltage equals the differential gain, A_d. Similarly, if $V_+ = V_- = 1$ V, then $V_d = 0$ and $V_c = 1$, and the output equals the common mode gain, A_c.

Usually, we are interested in having OAs with values of A_d and A_c as high and as low as possible, respectively. An ideal OA will present an infinite differential gain and a zero common mode gain. The common mode rejection ratio (CMRR) is given by

$$\beta = \left| A_d/A_c \right|. \tag{2.38}$$

From Equation (2.37) it is deduced that

$$V_o = A_d V_d \left(1 + \frac{1}{\beta} \frac{V_c}{V_d} \right), \tag{2.39}$$

which shows how the OA response approaches ideality when the CMRR is very high.

OAs can also be operated with ac signals; however, the differential gain decreases as the frequency increases. A typical gain–frequency curve of a $\mu741$ operated in open-loop mode is shown in Figure 2.37. The response of

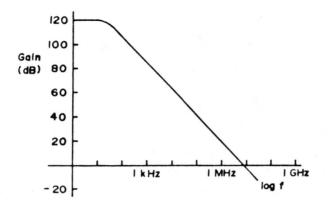

Figure 2.37. Bode plot or gain–frequency curve of an OA. The gain scale is given in decibels [dB = $20 \log(V_o/V_i)$].

an ideal OA would not be frequency dependent, and the initial plateau would be infinite. Real OAs have a finite working frequency band. The frequency that corresponds to half the gain in the open-loop working mode is called the transition frequency.

Another desirable feature of an ideal OA is a zero offset differential voltage; that is, the OA should give a zero output when the same voltage is applied to the inputs. However, in the inputs of real OAs, small voltages are always present. After amplification, these voltages can drive the output even to saturation. This problem is avoided by connecting a loop with a potentiometer to the offset null pins. However, in most applications, this is not necessary.

Finally, other characteristics for an ideal OA would be:

Infinite input impedance, Z_i. Real OAs reach values of several megohms.

Zero output impedance, Z_o. Typical real values are 100–200 Ω.

Zero input polarization current, I_b. Real values are on the order of the microamperes.

2.12.3. Inverting Amplifier

Application of rule 1 (see Section 2.12.1) to the circuit of Figure 2.38 leads to the conclusion that, since the potential at B is at ground, point A must also be at ground. It is said that point A is a virtual ground. Therefore, the voltage

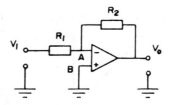

Figure 2.38. Inverting amplifier.

across R_2 is V_o and the voltage across R_1 is V_i. Also, from rule 2 it is deduced that

$$I_o = -I_i,\qquad(2.40)$$

and therefore

$$\frac{V_o}{R_2} = -\frac{V_i}{R_1}.\qquad(2.41)$$

Rearranging, we obtain

$$G = \frac{V_o}{V_i} = -\frac{R_2}{R_1},\qquad(2.42)$$

which is the gain of the amplifier (not the gain of the OA, but that of the amplifying circuit). Since point A is always at zero volts, the input impedance is $Z_1 = R_1$.

This analysis is true for ac and dc, which means that this circuit can amplify both types of voltage. If only the ac component of a mixed signal must be amplified, a blocking capacitor may be inserted in the input.

Usually, a third resistor is inserted between the noninverting input and the ground. Its resistance must fulfill the condition

$$\frac{1}{R_3} = \frac{1}{R_1} + \frac{1}{R_2};\qquad(2.43)$$

that is, R_3 equals the value of R_1 and R_2 when they are mounted in parallel, which is usually expressed as

$$R_3 = R_1 \| R_2.\qquad(2.44)$$

The function of R_3 is to equalize the small currents present at the inputs when the OA is standing by.

2.12.4. Noninverting Amplifier

Application of rule 1 to the circuit in Figure 2.39 gives $V_A = V_i$. On the other hand, the inverting input (A) is part of a voltage divider. From Equation (2.3) we have

$$V_A = V_o \frac{R_1}{R_1 + R_2}, \tag{2.45}$$

and therefore

$$G = \frac{V_o}{V_i} = 1 + \frac{R_2}{R_1}, \tag{2.46}$$

which is the gain. The input impedance depends on the OA and is very high, typically on the order of $10^8 \, \Omega$ for a μA741, and even more than $10^{12} \, \Omega$ if a FET input OA is used instead.

2.12.5. OA-Based Voltage Follower

The OA equivalent of a transistor-based emitter–follower is shown in Figure 2.40. In fact, this is a noninverting amplifier but with an infinite value of the resistance between inverting input and ground, R_1. From Equation (2.46) we deduce that the gain is unity. Thus, this circuit is just a follower. Followers are used as a first stage in many circuits, when a high input impedance is required. An amplifier of gain unity is also called a buffer, because of the high input and low output impedances.

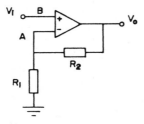

Figure 2.39. Noninverting amplifier. As is shown later, this circuit can work as a current source.

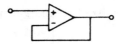

Figure 2.40. An OA working as voltage follower.

2.12.6.   OA-Based Current Sources

The circuit of Figure 2.39 approximates an ideal current source. Since no current is drawn by the inputs and the potential at the positive and negative inputs is the same, we have

$$I_{\text{load}} = I_{R_1} = \frac{V_i}{R_1} \tag{2.47}$$

which means that the current, flowing from the ground and through the load (R_2) is constant as long as V_i is maintained constant.

The major disadvantage of this circuit is that the load is not grounded. A simple, quality current source, provided with ground for the load, can be constructed with an OA and an auxiliary *pnp* transistor. The circuit is shown in Figure 2.41. The voltage divider on the left sets the potential of the inputs and the potential drop across R. The latter, $V_s - V_{R_2}$, forces a current through the emitter–collector path and the load, which is given by

$$I_{\text{load}} = \frac{V_s - V_{R_2}}{R}. \tag{2.48}$$

The output current, which drives the transistor, is proportional to the voltage drop applied to the OA noninverting input, V_{R_2}. A variable current source is obtained by using a potentiometer instead of a fixed resistor at R_2.

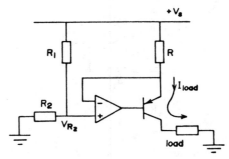

Figure 2.41. A quality current source.

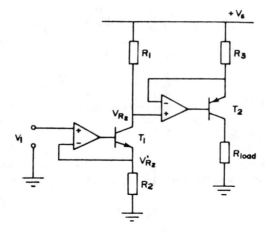

Figure 2.42. A voltage-driven current source.

A voltage driven current source is obtained by using an auxiliary circuit, as shown in Figure 2.42. The auxiliary circuit is an emitter–follower based on an *npn* transistor. The external input voltage regulates the voltage at V'_{R_2} and therefore also at V_{R_2}.

2.12.7. Current-to-Voltage Conversion

A single resistor mounted as shown in Figure 2.43 constitutes the simplest current-to-voltage (I/V) converter that can be imagined. In spite of its simplicity, this is a useful procedure to measure a current with an instrument designed to measure voltages, such as a paper chart recorder. However, this approach has the disadvantage of presenting a nonzero impedance to the source of input current. This means that the signal will depend on the load, that is, on the value of R and on the input impedance of the instrument connected to V_0.

A better I/V converter built up with an OA is shown in Figure 2.44. Here the circuit converts the small photocurrent generated by a photodiode into

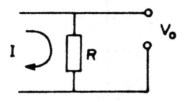

Figure 2.43. The simplest I/V converter.

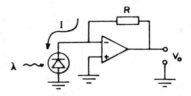

Figure 2.44. An OA-based I/V converter used as a high-input impedance preamplification step for a photodiode output.

an easily measurable voltage. The inverting input is a virtual ground and, since no current flows through the inputs, the output voltage is given by

$$V_o = RI, \tag{2.49}$$

which means that it is proportional to the photocurrent. If $R = 1\ \text{M}\Omega$, 1 V/ μA will be obtained. Although not shown in the figure, connection of the noninverting input to the ground is usually done through a resistor.

2.12.8.   Summing Amplifier

The circuit shown in Figure 2.45 is basically an inverting amplifier, where the input current is given by

$$I_i = I_o = \frac{V_o}{R_o} = \sum_{k=1}^{n} I_k = \frac{V_1}{R_1} + \frac{V_2}{R_2} + \frac{V_3}{R_3} + \frac{V_4}{R_4}. \tag{2.50}$$

The output voltage is therefore a weighted sum of voltages. Let us assume that the voltages are highs and lows of a family of logics, for example, +5 V and 0 V for TTL levels, and that they represent the first, second, third and fourth binary digits. By using resistors that are proportional to R, $2R$, $4R$,

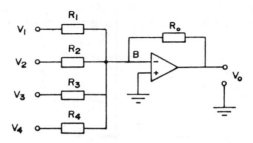

Figure 2.45. A simple summing amplifier for D/A conversion.

and $8R$, the output in volts will be proportional to the binary number. From equation (2.50) we can write:

$$V_o = \frac{R_o}{R} \left(V_1 + \frac{V_2}{2} + \frac{V_3}{4} + \frac{V_4}{8} \right) \qquad (2.51)$$

In this way, a digital number formed by a combination of highs and lows gives a stepped analog output. This is the basis of a popular technique for digital-to-analog conversion (D/A conversion), although D/A converters contain more complex circuits, including a larger number of resistors.

2.12.9.   OA-Based Integrators and Differentiators

Nearly perfect integrators, without the restriction $V_o \ll V_i$ (see Section 2.6.3), can be constructed with OAs. In the circuit of Figure 2.46, the input current, flowing through C, is given by

$$I_i = \frac{V_i}{R}. \qquad (2.52)$$

Since the inverting input is a virtual ground, the output voltage is given by

$$\frac{V_i}{R} = -C \frac{dV_o}{dt}. \qquad (2.53)$$

Integration gives

$$V_o = - \frac{1}{RC} \int_0^t V_i dt + K. \qquad (2.54)$$

When the input is a square wave, the voltage is constant along each half cycle and the output is a triangle wave:

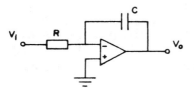

Figure 2.46. An OA-based integrator.

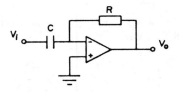

Figure 2.47. An OA-based differentiator.

$$V_o = - \frac{1}{RC} \int V_i dt = - \frac{V}{RC} t + K. \qquad (2.55)$$

On the other hand, if R and C are reversed as shown in Figure 2.47, the resulting circuit is a differentiator. Since the inverting input is at ground, the rate of change of input voltage produces a current that is given by

$$IR = C \frac{dV_i}{dt} = V_o. \qquad (2.56)$$

If the input is a triangle wave, the derivative, dV_i/dt, will be a constant with sharp changes of sign, and therefore

$$V_o = -RC \frac{dV_i}{dt} = K; \qquad (2.57)$$

that is, a periodic square wave is obtained at the output. This circuit and a similar one for integration of a square wave are very useful for signal conversion in interfacing digital and analog circuits and in function generators.

2.12.10. Voltage Regulators Based on the 723 IC

The 723 chip is a regulable voltage regulator. It works with a floating ground, which permits one to obtain a variable output voltage ranging from 2 to 37 V. The internal components and pin connections are shown in Figure 2.48. Other specifications are: maximum input voltage, 40 V; maximum load current, 150 mA; reference voltage, $V_{ref} = 7.15$ V ($\pm 5\%$); zener voltage, $V_z = 6.2$ V ($\pm 5\%$); dissipated power, 900 mW.

A regulable voltage source built up with a 723 is shown in Figure 2.49. A regulated output voltage, smaller or higher than V_{ref} may be obtained with this circuit. An output voltage smaller than V_{ref} is achieved by making R_5

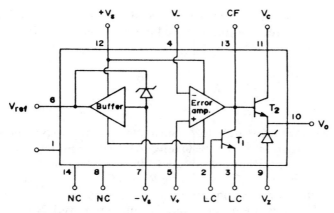

Figure 2.48. The 723 internal circuitry and external connections.

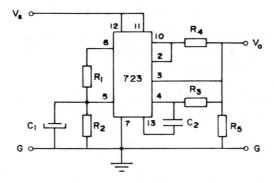

Figure 2.49. A 723-based regulable voltage source.

infinite (open) and with the help of the $R_1 - R_2$ voltage divider. The potential at the noninverting input of the internal OA is

$$V_+ = \frac{R_2}{R_2 + R_1} V_{\text{ref}}. \tag{2.58}$$

When the circuit is standing by, resistor R_3, with a value $R_3 = R_2 \| R_1$, maintains the same potential at the two OA inputs; that is:

$$V_0(\text{pin } 13) = V_+ = V_-. \tag{2.59}$$

The output current is limited by transistor T_1 (see Figure 2.48) and depends on the value of R_4 (connected between base and emitter, that is, at pins 2 and 3). Its value is

$$I_{o,max} = \frac{0.6}{R_4}. \tag{2.60}$$

Let us assume that we want to design a power supply for 5 V and 100 mA. The value of R_2 can arbitrarily be taken as 4K7. Using Equations (2.58) and (2.59), we have $R_1 = 2K$ and, from the voltage divider equation, $R_3 = 1K5$. Finally, from Equation (2.60) we have $R_4 = 6\ \Omega$.

On the other hand, to obtain an output higher than V_{ref}, the voltage divider on the left is not needed, and R_2 can be removed. Now we have

$$V_o = \frac{R_2 + R_3}{R_3} V_{ref}. \tag{2.61}$$

The values of the resistors for a given output can easily be found using Equations (2.59) and (2.61). The necessary values for the 5 V/100 mA and a 12 V/75 mA voltage source are given in Table 2.3.

However, the maximum current that can be obtained with these circuits (about 150 mA) is too small for many applications. Higher currents are obtained with an auxiliary transistor mounted with its base connected to the 723 output, as shown in Figure 2.50. In this circuit, the maximum output current is only limited by the type of transistor used.

Table 2.3. Component Specifications to Construct Two Regulable Voltage Sources Using the Scheme of Figure 2.49

Component	Values	
V_s	+8 V	+16 V
V_o	+5 V, 100 mA	+12 V, 75 mA
R_1	2K	1K2
R_2	4K7	1K5
R_3	1K5	0
R_4	6.8 Ω	8 Ω
R_5	∞	2K2
C_1	4.7 μF	4.7 μF
C_2	100 pF	100 pF

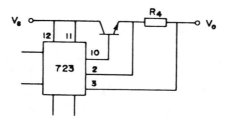

Figure 2.50. The 723 output drives an auxiliary transistor.

A complete circuit of a regulable voltage source that supplies a large value of power is shown in Figure 2.51. Output voltage is regulable from 2 to 15 V and maximum output current can be as high as 1 A. Input voltage is supplied by a full-wave bridge rectifier provided with a filter and an indicating LED. Output current is given by the TIP31A transistor. The 1N4002 diode, connected between emitter and collector, protects the circuit against excessive external voltages, which can accidentally be applied to the output.

Let us see how some of the components of this circuit may be calculated. For a minimum output of 2 V,

$$V_o = V_+ = 2 \text{ V} = \frac{R_2}{R_2 + R_1} V_{\text{ref}}. \qquad (2.62)$$

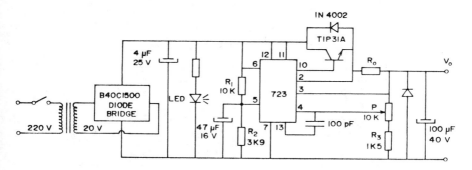

Figure 2.51. A regulable voltage source for large loads based on a 723 IC.

If a value R_2 = 3K9 is chosen, R_1 = 10K. The maximum output is obtained with the potentiometer P. Then

$$V_{o,max} = 15 \text{ V} = \frac{R_3 + P}{R_3} V_+ = \frac{R_3 + P}{R_3} 2 \text{ V}. \qquad (2.63)$$

If a value R_2 = 1K5 is chosen, then P = 10K. Finally, for a maximum load current of 1 A,

$$R_o = \frac{0.6}{1 \text{ A}} = 0.6 \text{ }\Omega. \qquad (2.64)$$

2.12.11.   Single Power Supply and Multiple OAs

It is not necessary to power an OA with a symmetrically split supply of a specified voltage, such as ±15 V for a μA741C. OAs can also be operated with lower and even asymmetric voltages, as long as the total supply voltage (voltage difference between the supply pins) is within specifications; that is, it does not exceed the maximum and reach the necessary minimum, such as 30 V and a few volts, respectively, for a μA741C.

Furthermore, there are many occasions where it is convenient to use a single supply, connected to the positive or to the negative supply pin, the other one being simply grounded. However, the usual technique for most applications is the symmetrically split supply.

Chips containing several OAs are also available. Thus, as shown in

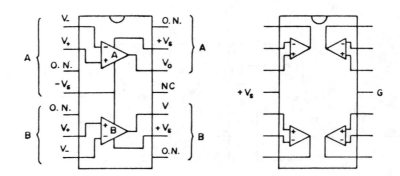

Figure 2.52. A dual and a quadruple OA. Some of the internal connections are shown. NC = not connected. O.N. = offset null.

Figure 2.52, the 747 is a dual OA; that is, the chip contains two OAs, each one of them being equivalent to one μA741. The internal structure of the quadruple LM324 (quad LM324) is also shown in the figure. This chip contains four high-gain, frequency-compensated OAs. The LM324 is especially designed to work with a single supply and with a wide range of supply voltages. Other OAs that also work with a single supply are the μA798 (dual), the μA799 (single with offset null), and the LM358 (dual).

2.12.12. BIFET and BIMOS Amplifiers

The characteristics of conventional OAs are far from those required for ideal OAs. Thus, for instance, the μA741 has an input impedance that is only on the order of 1 MΩ and dissipates a considerable amount of power. Better performances are obtained from the newer BIFET and BIMOS OAs.

BIFETs are ICs that combine bipolar transistors and FETs. The first BIFET OAs, the LF155 series, were commercialized by National Semiconductors in 1975. The main difference with conventional OAs is that the chip inputs are constituted by JFETs. The characteristics of two BIFET OAs are given and compared with those of the μA741 in Table 2.4.

BIMOS OAs combine bipolar and MOSFET transistors as input and output stages. The first BIMOS OA, the CA3130, was introduced by RCA, also in the mid 1970s. A similar circuit, the CA3140, is pin compatible with

Table 2.4. Comparison of the 741 OA with Two Typical BIFET OAs, the LF356 and LF357

Characteristics	741	LF356	LF357
Supply range	±2 to ±18 V	±3 to ±18V	±3 to ±18 V
Maximum output load	25 mA	15 mA	15 mA
Input polarization current	200 μA	30 pA	30 pA
Output voltage offset	±1 mV	±3 mV	±3 mV
Input impedance	1 MΩ	1×10^6 MΩ	1×10^6 MΩ
Gain (without load)	100 dB	106 dB	106 dB
CMRR	90 dB	100 dB	100 dB
Transition frequency	1 MHz	5 MHz	20 MHz

the μA741; that is, it can be used as a pin-to-pin substitute of the μA741. However, in a CA3140, only the inputs have MOSFETs, all other on-chip transistors being bipolar.

2.13. COMPARATORS

2.13.1. Introduction

Comparing two signals, for example, to know which of them is the largest, or comparing a signal with a preset value is done with a comparator. A practical example is found in the generation of a triangle wave. A current is supplied to a capacitor and the output voltage increases. The comparator detects when the desired peak voltage is reached for the polarity of the current to be reversed. A comparator is also essential in analog-to-digital conversion (A/D conversion) in measurement instruments, such as digital millivoltmeters or pH meters.

To convert a voltage to a number, the unknown voltage is applied to one of the inputs of a comparator, and a linear ramp is applied to the other input. As long as the ramp voltage is smaller than the unknown, a digital counter counts cycles of an oscillator. When both signals are equal, the result of the count is displayed. This process is called single-slope integration.

A very simple comparator consists of a single high-gain differential amplifier working in open-loop configuration, as shown in Figure 2.53. Since the voltage gain of the open-loop circuit is higher than 10^5, less than 1 mV of difference between the inputs is enough to saturate the output. Thus, when the voltages applied to the inputs approach each other, equalize, and finally cross each other, the output goes from positive to negative saturation or vice versa.

ICs especially designed to be used as comparators, such as the LM311, are also available. An advantage of these circuits, in comparison with a simple OA, is their fast response. Furthermore, in an OA, the output ranges within some limits which are given by the supply voltages, for example, approximately ± 13 V for a μA741C supplied with ± 15 V. Instead, the output of an IC comparator is an open collector that is supplied with the desired saturation voltage.

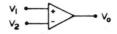

Figure 2.53. The simplest comparator.

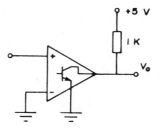

Figure 2.54. A comparator used to convert an analog input into any of the two TTL logic levels.

On the other hand, as shown in Figure 2.54, the emitter is grounded. The output goes from ground to the applied voltage when the input signal becomes positive and vice versa. In the circuit of Figure 2.54, the applied voltage is the TTL high, +5 V. Therefore, this is a simple example of conversion of an analog signal, the input, into a logic level. Comparators are very useful to perform A/D conversions.

2.13.2. The Shmitt Trigger

The Shmitt trigger is a regenerative comparator. It suppresses the effects of noise, which causes the output of a simple comparator to oscillate when the input approaches the transition voltage, and is based on an OA with positive feedback. The circuit and hysteresis cycle are shown in Figure 2.55. Hysteresis means that a "dead zone" region is created between the comparator states. Peak and valley potentials, V_p and V_v, applied to the inverting input, control the sign of the output.

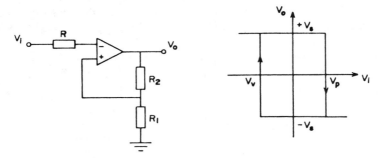

Figure 2.55. The Shmitt trigger: scheme and hysteresis cycle.

For a minimum offset error, the value of R must be $R = R_1 \| R_2$. At the upper part of the cycle, V_o is at positive saturation; that is, $V_o \approx +V_s$. The potential at the noninverting input is given by the voltage divider:

$$V_+ = \frac{V_s R_1}{R_1 + R_2}.$$ (2.65)

If the output is required to change its sign, a peak voltage V_p should be applied to the inverting input. The threshold value for V_p is given by

$$V_p > V_+ = \frac{V_o R_1}{R_1 + R_2}.$$ (2.66)

The output is now $V_o \approx -V_s$, and

$$V_+ = \frac{-V_s R_1}{R_1 + R_2}.$$ (2.67)

To get back to the initial state, a valley voltage must be applied. Its threshold value is

$$V_v < V_- = \frac{V_o R_1}{R_1 + R_2}.$$ (2.68)

This is also a useful circuit for analog-to-digital conversion.

2.13.3. A Window Comparator

The circuit of Figure 2.56 has two outputs, $V_{o.1}$ and $V_{o.2}$, which correspond to two OAs, A_1 and A_2, with a single common input, V_i. Both OAs are mounted as voltage comparators and, since they are working in an open loop, the outputs are in positive or negative saturation.

The noninverting inputs are at the reference voltages, $V_{1(+)}$ for A_1 and $V_{2(+)}$ for A_2. Depending on the value of the input V_i, the following cases are possible:

1. $V_i < V_{2(+)} < V_{1(+)}$, then $V_{o.1} = +V_s$ and $V_{o.2} = +V_s$
2. $V_{2(+)} < V_i < V_{1(+)}$, then $V_{o.1} = +V_s$ and $V_{o.2} = -V_s$
3. $V_{2(+)} < V_{1(+)} < V_i$, then $V_{o.1} = -V_s$ and $V_{o.2} = -V_s$

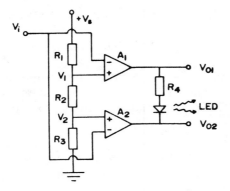

Figure 2.56. A window comparator.

When the input fulfills the conditions of case 2, it is said to be within the range or "window." The desired window limits are set by selecting the resistances of the voltage divider.

Let us assume that we want to establish the range +5.0–5.5 V for the window and that the power supply is +12 V. Only when the input is within the window is there a nonzero voltage drop across the two outputs, and the LED is turned on. If we set the value of the current through the voltage divider to 1 mA, the resistances must be

$$R_1 = \frac{V_+ - V_{1(+)}}{I} = \frac{12 - 5.5}{0.001} = 6K5, \tag{2.69}$$

$$R_2 = \frac{V_{1(+)} - V_{2(+)}}{I} = \frac{5.5 - 5.0}{0.001} = 500 \ \Omega, \tag{2.70}$$

$$R_3 = \frac{V_{2(+)}}{I} = \frac{5.0}{0.001} = 5K. \tag{2.71}$$

On the other hand, if the LED requires a current of about 15 mA to light properly, and if the typical voltage drop across it is 1 V, then

$$R_4 = \frac{2 \times 12 - 1}{0.015} = 1K5. \tag{2.72}$$

2.14. DIFFERENTIAL AMPLIFIERS

2.14.1. Introduction

A differential amplifier is an electronic device that produces an output voltage that is the product of the gain and the difference between two input voltages. Figure 2.57 shows a simple dc differential amplifier, whose gain is R_2/R_1:

$$V_{out} = (V_2 - V_1) \frac{R_2}{R_1}. \qquad (2.73)$$

The potential difference $V_2 - V_1$ is known as the differential mode signal, and the differential amplifier ideally responds only to such signals. A common mode signal is that which is applied to both inputs simultaneously. In the ideal differential amplifier the common mode gain is zero.

Instrumentation applications, including many in the data-acquisition field, are ideal for differential amplifiers. In the circuit of Figure 2.58, a differential amplifier is used to amplify the differential voltage given by a Wheatstone bridge (see Section 5.3.2). The R_1 and R_2 resistors are the passive arm of the bridge, whereas P_1, R_3, and NTC are the active arm. The NTC is a thermistor, a temperature-sensitive resistor. Obviously, a similar variable resistor sensitive to any other physical parameter can be used instead.

The voltage applied to the bridge is given by the reference zener and its value is 6.8 V. Thus, between R_1 and R_2, the voltage is ≈ 3.4 V. A regulable thermometer is built up by connecting an OA, wired in the differential amplifier mode, at the active and passive arm corners of the bridge ($R_3 - R_4$ and $R_1 - R_2$, respectively) and measuring the output voltage.

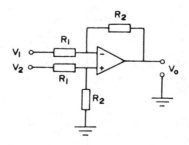

Figure 2.57. A simple differential amplifier.

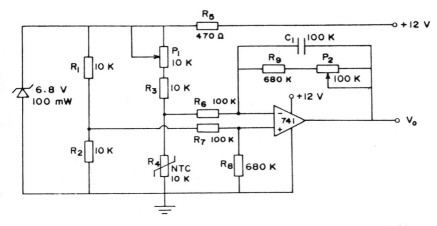

Figure 2.58. A differential amplifier measuring the response of a Wheatstone bridge.

Calibration of the thermometer, for example, to the Celsius scale, is performed by submerging the thermistor in a melting ice bath and adjusting the P_1 potentiometer until the output voltage reaches zero. This would correspond to a voltage of 3.4 V at the active bridge corner.

Next, the thermistor is submerged in a boiling water bath and the gain of the differential OA is adjusted to 100 by means of the feedback potentiometer P_2. Hereforth, the output, measured with a voltmeter, will give the temperature of the thermistor in degrees Celsius. However, since a thermistor response is nonlinear, the output will not give a linear temperature scale either. A much better linear response can be obtained as explained in Section 5.3.2.

2.14.2. Instrumentation Amplifiers

Especially designed dc-coupled differential amplifiers are frequently needed to measure the small differential signals coming from transducers. A standard instrumentation amplifier is shown in Figure 2.59. The dual OA input stage provides a high input impedance, a high differential gain, and unity common mode gain. Thus, a high common mode rejection ratio (CMRR) is achieved. The differential gain is given by

$$G = 1 + \frac{2R_2}{R_1}. \tag{2.74}$$

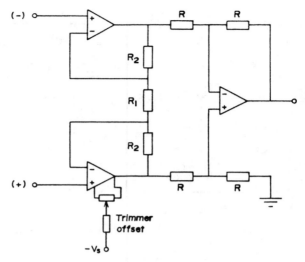

Figure 2.59. A standard instrumentation amplifier.

The differential input of this first stage is used to drive a conventional differential amplifier circuit. The latter is frequently designed for unity gain and used to generate the required single output, as well as to improve the CMRR by further reduction of the remaining common mode signal.

IC instrumentation amplifiers containing this standard configuration are available from several manufacturers. All components are internal, except the resistor R_1, which can be changed to set the gain at the desired value.

Popular instrumentation amplifiers are the LM-0036, the AD522, and the 3630. The LM-0036 can be supplied with voltages as low as ± 1 V. The 3630 is a high-accuracy amplifier, with a gain linearity better than 0.002%. The gain of these ICs ranges from 1 to 100, and their input impedances are higher than 100 MΩ.

A complete instrumentation amplifier circuit is shown in Figure 2.60. An auxiliary OA (G) is used as a "guard" to reduce the effects of cable capacitance and input leakage. The output of the guard is connected to the input cable shield. The sensing and reference terminals provide positive and negative feedback from the output voltage at the load to the amplifier inputs. The gain is

$$G = \left(1 + \frac{2R_2}{R_1}\right)\frac{R_4}{R_3}. \tag{2.75}$$

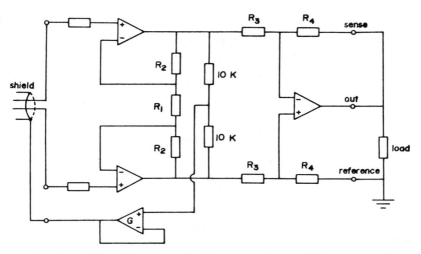

Figure 2.60. A complete instrumentation amplifier.

Several inexpensive IC instrumentation amplifiers are available from several manufacturers. Thus, for instance, the differential transconductance amplifiers, such as the LF352 and the AD521, achieve high CMRR without the need of matched external resistors. The gain is set by the ratio of a pair of external resistors. The National Semiconductor LM-363-XX family (Figure 2.61) includes a number of devices with fixed or switch-selectable gain, where XX indicates the gain, which can be 10, 100, or 500.

The LM-363 AD is a 16-pin miniDIP chip that provides step-, selectable fixed gains of 10, 100, or 1000, which are selected by connecting pins 2

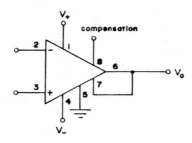

Figure 2.61. The LM-363-XX fixed-gain package.

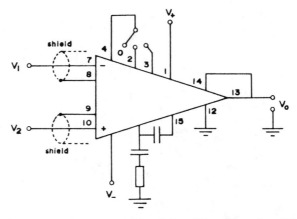

Figure 2.62. The LM-363 AD. Switch at positions 0, 2, and 3 provides gains of 10, 100, and 1000, respectively. The capacitors are optional, for frequency compensation.

or 3 with 4 (Figure 2.62). The circuit uses guard shield driver terminals, which are highly recommended when large gains are needed.

Very simple instrumentation amplifier circuits can be constructed with the Precision Monolithic AMP-01 IC (Figure 2.63). The gain is given by $G = 20R_2/R_1$, with a permissible range within 0.1–10,000.

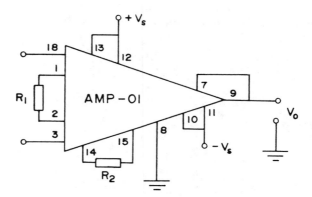

Figure 2.63. The Precision Monolithic AMP-01.

2.15. ISOLATION AMPLIFIERS

In some applications, the signal input must be electrically isolated from the instrumentation, either for safety or to avoid noise, when small differences between two voltages have to be amplified.

A method is to use an isolation amplifier, which is represented in Figure 2.64. The first stage of the amplifier drives a LED, which lights a matched pair of photodiodes. These constitute the inputs of the isolated second stage. Isolation amplifiers require the use of separate power supplies for the two stages. The isolated power supply may be a battery, whereas the non-isolated power supply terminals are connected to the usual dc supply. A 722 dual dc-to-dc converter may also be used. This device produces two independent ±15 V dc supplies, which are isolated from the power mains and from each other.

A second method is based on the use of a transformer. The dc signal controls the amplitude of an oscillator (see below), and the modulated signal feeds the primary winding of the transformer.

2.16. OSCILLATORS

Most electronic instruments contain oscillators or waveform generators. Oscillators are used not only as generators of periodic signals and pulses but also to control periodic states or any time-dependent operation. An oscillator can be used simply as a source of regularly spaced pulses, for example, as a clock or as a time base for a frequency counter. Thus, oscillators are used in oscilloscopes, radiofrequency receivers, computers and computer peripherals, and in nearly every digital instrument such as timers, counters, calculators, and multimeters.

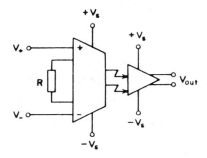

Figure 2.64. Isolation amplifier. The gain is settled by the value of R.

A simple oscillator can be built up with a ramp generator (capacitor + resistor) and a comparator. The capacitor charge increases until the voltage reaches a preset value. At this point, the comparator output triggers the discharge of the capacitor, and the cycle begins again. In this way, a sawtooth is produced. Alternatively, the polarity of the charging current can be reversed, and a triangle wave is generated. This type of generator is called a relaxation oscillator and, if properly designed, it can generate waveforms that are very stable in frequency.

The circuit of a RC relaxation oscillator is shown in Figure 2.65. When power is applied, the output goes to positive saturation and the capacitor begins to charge toward $+V_s$. The rate of the charging process is controlled by the time constant, RC. When the voltage reaches the point where $V_i = +V_s/2$, the OA swings into negative saturation, and the capacitor begins discharging toward $-V_s$. An approximately square-wave signal, with an amplitude equal to $+V_s-(-V_s)$ and a frequency of $2.2RC$ is obtained.

A relaxation oscillator can also be constructed with the timer IC known as 555. The circuit is shown in Figure 2.66. The 555 operation is controlled by the low threshold or trigger, and the high threshold inputs (pins 2 and 6, respectively). The trigger is activated by a voltage below $V_{cc}/3$, V_{cc} being the voltage at pin 8. Similarly, the threshold is activated by an input level above $2V_{cc}/3$. When the trigger is activated, the output goes high and remains there until the threshold is also activated. Then, the output goes low and the discharge path, which goes from pin 7 to pin 1, is switched on.

When power is applied, the capacitor begins charging up and the voltage also goes up, activating the trigger and thus giving a high logic level at the output. This process continues until the threshold is reached. Then, the output goes low, and the capacitor discharges, until the trigger voltage is reached. Next, the discharging path is switched off and the cycle begins again. The output is a square wave with a period given by

$$T = 0.693(R_A + 2R_B)C. \qquad (2.76)$$

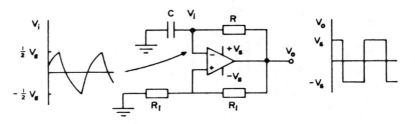

Figure 2.65. A RC relaxation oscillator.

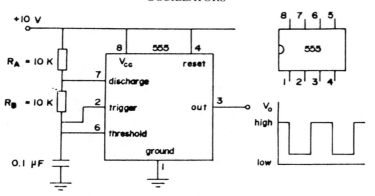

Figure 2.66. A relaxation oscillator based on the 555 timer circuit.

The amplitude, which depends on V_{cc}, can be regulated from 4.5 to 16 V. Frequency stability is about 1% and is only slightly affected by variations of the supply voltage, because the low and high thresholds track the supply fluctuations. In Chapter 3, we show how this and other circuits based on the 555 timer IC are frequently used in digital electronics.

When a highly stabilized frequency is required, a crystal oscillator should be used. Crystal oscillators are based on a piece of quartz that vibrates at a given highly constant frequency.

CHAPTER

3

DIGITAL ELECTRONICS

3.1. INTRODUCTION

Digital electronics deals with electronic digital circuits, which handle digital or binary encoded information. In digital electronics, information is digitalized to be stored and processed by using the logic of Boole, based on the existence of only two possible logic states, the logic 0 and the logic 1, or "true" and "false." These states are represented by two voltages, a high and a low (such as +5 V and ground for the TTL family), or by the open or closed states of a switch.

Digital circuits are important because of how simply high-speed switching devices can be constructed. Changes of state, from high to low or vice versa, can be made in the nanosecond range, which allows handling of huge amounts of information in a short time. In this chapter, the basic concepts of digital electronics, as well as some useful digital circuits, are presented.

3.2. BITS, BYTES, WORDS, AND DIGITAL SIGNALS

The bit, acronym for binary digit, is the elemental unit of information. One bit is equivalent to a single decision between the two possible logic states, 0 or 1. A bit series represents information; for example, the number 8 is encoded as the 4-bit number 1000, the letter "A" in the ASCII code is encoded as the 8-bit number 11000001. A group of 8 bits is called a byte, and the number of bits that can be handled simultaneously by a computer is a "word." The term byte has become popular since many computers work with words that are composed of multiples of 8 bits, such as 8-, 16-, or 32-bit words.

Digital signals are usually voltages that follow the requirements of one of the families of digital logics. Digital signals can adopt the form of a steady voltage, a change of voltage such as an upward or a downward edge, a single pulse, a group of pulses, or a square wave.

3.3. THE BINARY, OCTAL, AND HEXADECIMAL CODES

The simplest way to code a number is to use the base "2." Only the decimal 0 and 1 can be represented by the binary numbers 0 and 1. Translation of higher decimal numbers into binary and vice versa requires a calculation. Thus, for instance,

$$1111_2 = 1\times2^3 + 1\times2^2 + 1\times2^1 + 1\times2^0 = 8 + 4 + 2 + 1 = 15_{10},$$
$$1001_2 = 1\times2^3 + 1\times2^0 = 8 + 1 = 9_{10}.$$

Numbers expressed in other bases, as in octal (base 8) and hexadecimal (base 16), are also frequently used. The equivalents of the 16 first decimal numbers, from 0 to 15, as well as some other important numbers, in decimal, binary, and hexadecimal (hex), are given in Table 3.1 and 3.2. An easy way of translating a decimal number to octal or to hex with a computer is to use the BASIC functions V\$ = OCT\$(x) and V\$ = HEX\$(X).

In some applications it is necessary to use some numbers that can represent, for instance, microprocessor instructions. When expressed in binary code, large numbers are difficult to remember and to use without typing errors. It is much more convenient to express these numbers in hexa-

Table 3.1. Equivalents of the First Natural Numbers in Usual Encoding Bases

Decimal	Binary	Hex	Octal
0	0000	0	0
1	0001	1	1
2	0010	2	2
3	0011	3	3
4	0100	4	4
5	0101	5	5
6	0110	6	6
7	0111	7	7
8	1000	8	10
9	1001	9	11
10	1010	A	12
11	1011	B	13
12	1100	C	14
13	1101	D	15
14	1110	E	16
15	1111	F	17

Table 3.2. Equivalents of Some Important Numbers

Decimal	Binary	Hex
0	0000	0
15	1111	F
255	1111 1111	FF
1023	11 1111 1111	3FF
4095	1111 1111 1111	FFF
65535	1111 1111 1111 1111	FFFF

decimal. For example, the hex numbers 32 and 3A represent instructions for the Intel 8085 microprocessor. The binary equivalents and meaning are:

Binary	Hex	Mnemonic Instruction and Meaning
0011 0010	32	STA, store in register A
0011 1010	3A	LDA, retrieve (load) from register A

Since the highest value of a digit in hex is F, which corresponds to the highest 4-digit binary number, 1111, it is easy to transform a binary into a hex number: the former is divided into packets or "nibbles," each one containing 4 digits. Then, each nibble is translated to hex independently. Thus, for instance, a binary number having from 5 to 8 digits is divided into two 4-digit nibbles, after adding the necessary zeros to the left of the binary number, and a hex digit per nibble is obtained. For example,

$$110001_2 = 0011\ 0001 = 31_{16}.$$

The simplicity of translation is one of the reasons of using base 16.

3.4. OTHER DIGITAL CODES

Because digital circuits work on the basis of two logic states, all arithmetic operations must be performed with binary numbers. However, we are used to working with decimal numbers, and therefore the usual way of working with microprocessors is to introduce the data in decimal form. Data are then codified in the binary system, operations are performed in this form, and, finally, the results are decodified from binary to decimal.

Thus, encoders and decoders are found in computers as well as in laboratory measuring instruments. In the latter, analog information is transformed directly into binary form, processed, and, finally, decoded to decimal.

Several types of digital or binary code are used by microprocessors and microprocessor-based devices, such as computers. Thus, calculations are frequently performed by using the binary-coded decimal or BCD code. In this popular code, each decimal digit is translated individually into the corresponding 4-bit binary number. For example,

$$207_{10} = 0010\ 0000\ 0111.$$

Other digital codes are correspondences between binary numbers and symbols, such as alphanumeric characters, arithmetic operators such as $+$, $-$, $\times$, and $\div$, other symbols such as % or ♥, as well as control functions, such as carriage return (CR) or line feed (LF). Examples are the ASCII and the EBCDIC codes. The ASCII code, or American Standard Code for Information Interchange, is a 7-bit code that includes the $2^7 = 128$ symbols represented in Table 3.3. The expanded 8-bit ASCII code contains 128 additional symbols, such as greek symbols and drawing elements.

Digital codes are also series of instructions used to perform operation sequences, for example, the instruction codes of the 8085 and the 8086 microprocessors. The binary expression corresponding to an instruction of a microprocessor is called a machine code, and the entire machine code of a microprocessor constitutes the machine language.

3.5. GATES

3.5.1. Introduction

In digital electronics, a gate is a circuit that has two or more inputs and an output. A digital datum obtained at the output is the result of a combination of the digital data applied to the inputs.

Gates are essential parts of the following devices:

Gated circuits, which control the transmission of digital signals.

Logic circuits, which provide a boolean logic relationship between input and output.

Memory elements, where logic zeros and ones, in the form of individual bits or groups of bits, are stored in, or retrieved from.

Table 3.3. The 7-Bit ASCII Code

Character	Decimal[a]	Hex	Description	Character	Decimal	Hex	Character	Decimal	Hex	Character	Decimal	Hex
^@	0	0	Null		32	20	@	64	40	`	96	60
^A	1	1	Start of heading	!	33	21	A	65	41	a	97	61
^B	2	2	Start of text	"	34	22	B	66	42	b	98	62
^C	3	3	End of text	#	35	23	C	67	43	c	99	63
^D	4	4	End of transmission	$	36	24	D	68	44	d	100	64
^E	5	5	Enquiry	%	37	25	E	69	45	e	101	65
^F	6	6	Acknowledge	&	38	26	F	70	46	f	102	66
^G	7	7	Bell	'	39	27	G	71	47	g	103	67
^H	8	8	Backspace	(	40	28	H	72	48	h	104	68
^I	9	9	Tab	)	41	29	I	73	49	i	105	69
^J	10	A	Line feed	*	42	2A	J	74	4A	j	106	6A
^K	11	B	Vertical tab	+	43	2B	K	75	4B	k	107	6B
^L	12	C	Form feed	,	44	2C	L	76	4C	l	108	6C
^M	13	D	Carriage return	-	45	2D	M	77	4D	m	109	6D
^N	14	E	Shift out	.	46	2E	N	78	4E	n	110	6E
^O	15	F	Shift in	/	47	2F	O	79	4F	o	111	6F
^P	16	10	Data link escape	0	48	30	P	80	50	p	112	70
^Q	17	11	Device control 1	1	49	31	Q	81	51	q	113	71
^R	18	12	Device control 2	2	50	32	R	82	52	r	114	72
^S	19	13	Device control 3	3	51	33	S	83	53	s	115	73
^T	20	14	Device control 4	4	52	34	T	84	54	t	116	74
^U	21	15	Negative acknowledge	5	53	35	U	85	55	u	117	75
^V	22	16	Synchronous idle	6	54	36	V	86	56	v	118	76
^W	23	17	Block	7	55	37	W	87	57	w	119	77
^X	24	18	Cancel	8	56	38	X	88	58	x	120	78
^Y	25	19	End of medium	9	57	39	Y	89	59	y	121	79
^Z	26	1A	Substitute	:	58	3A	Z	90	5A	z	122	7A
	27	1B	Escape	;	59	3B	[	91	5B	{	123	7B
	28	1C	File separator	<	60	3C	\	92	5C	|	124	7C
	29	1D	Group separator	=	61	3D	]	93	5D	}	125	7D
	30	1E	Record separator	>	62	3E	^	94	5E	≈	126	7E
	31	1F	Unit separator	?	63	3F	_	95	5F	DEL	127	7F

[a] Decimal values smaller than 27 correspond to control characters and greater than 31 to printed characters. Many of the nonprinting characters describe functions that are not usually incorporated into small computers.

69

The four common gates are called AND, OR, NAND, and NOR. A somewhat more complex gate is the exclusive OR or XOR. The logic function performed by a gate and, in general, by any digital circuit can easily be described by a "truth table." A truth table contains the output logic levels of a digital circuit which can be obtained from all the possible combinations of the input levels. Thus, a truth table gives a simplified and convenient picture of how a given digital circuit works.

3.5.2. The AND and NAND Gates

The symbol and truth table for a two-input AND gate are given in Figure 3.1. This gate requires that for the output to be at logic 1, both inputs must also be at 1. The scheme of circuit 7408, which contains four AND gates, is shown in Figure 3.2. The circuits of the 74XX series are provided with a supply connection for $+5.00 \pm 0.25$ V (pin 14) and a ground (pin 7). Care should be taken not to reverse these connections; otherwise the chip will probably be destroyed.

B	A	Q
0	0	0
0	1	0
1	0	0
1	1	1

Figure 3.1. The AND gate.

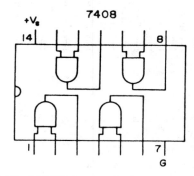

Figure 3.2. The 7408 IC showing the four internal AND gates.

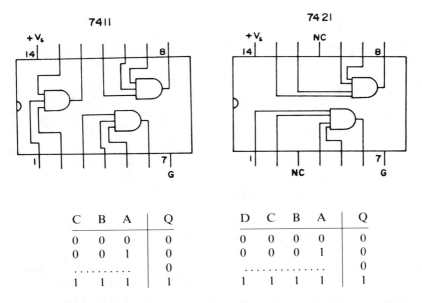

Figure 3.3. The 7411 and 7421 ICs. In the truth tables, the dotted lines stand for all the bit combinations not shown.

Other AND gates have a higher number of inputs, as the 7411 with three and the 7421 with four. The schemes and simplified truth tables are given in Figure 3.3.

The NAND gate is similar to the AND gate, but the logic output is reversed. This is represented by a small circle at the output, as shown in Figure 3.4. The 7400, 7410, 7420, and 7430 ICs have four 2-input, three 3-input, two 4-input, and one 8-input NAND gates, respectively. The schemes of the 7400 and 7430 ICs are given in Figure 3.5.

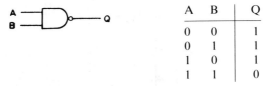

Figure 3.4. The NAND gate.

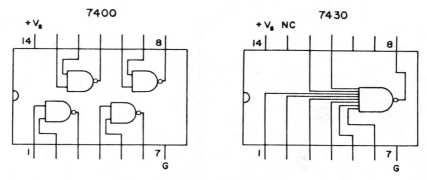

Figure 3.5. ICs containing NAND gates.

3.5.3.   Inverters and Separators

An inverter is a circuit with a single input and a single output that always have opposite logic states; that is, the output is at 1 when the input is at 0, and vice versa (Figure 3.6). The common 7404 IC (Figure 3.7) contains six internal inverters. As shown in Figure 3.8, a NAND gate can be constructed with an AND gate and an inverter, and vice versa.

A	Q
0	1
1	0

Figure 3.6. An inverter.

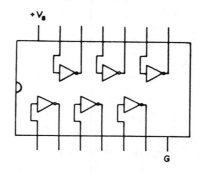

Figure 3.7. The 7404 IC showing the six internal inverters.

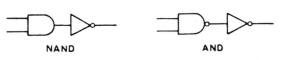

NAND   **AND**

Figure 3.8. Construction of NAND and AND gates with inverters.

On the other hand, in a separator, excitator, or buffer, the single input and output are always at the same logic level (Figure 3.9). Separators are useful in increasing the power along the paths of a digital circuit.

A	Q
0	0
1	1

Figure 3.9. A buffer.

3.5.4. The OR, NOR, and XOR Gates

For an OR gate, it is enough that one of the inputs be at high, for the output also to be at high. NOR gates are similar but with a reversed output. The symbols and truth tables of the OR and NOR gates are given in Figure 3.10. The 7432 IC, which contains four OR gates, is shown in Figure 3.11. The exclusive OR, or XOR gate, is similar to the OR gate, but when both inputs are equal the output is at low (Figure 3.12).

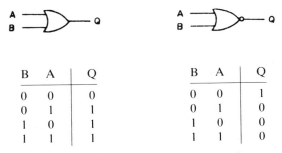

B	A	Q
0	0	0
0	1	1
1	0	1
1	1	1

B	A	Q
0	0	1
0	1	0
1	0	0
1	1	0

Figure 3.10. OR and NOR gates.

7432

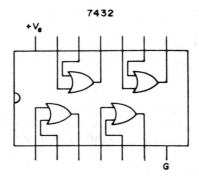

Figure 3.11. The 7432 IC.

B	A	Q
0	0	0
0	1	1
1	0	1
1	1	0

Figure 3.12. The XOR gate.

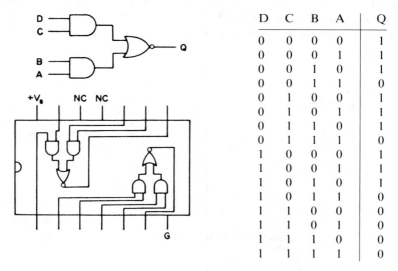

D	C	B	A	Q
0	0	0	0	1
0	0	0	1	1
0	0	1	0	1
0	0	1	1	0
0	1	0	0	1
0	1	0	1	1
0	1	1	0	1
0	1	1	1	0
1	0	0	0	1
1	0	0	1	1
1	0	1	0	1
1	0	1	1	0
1	1	0	0	0
1	1	0	1	0
1	1	1	0	0
1	1	1	1	0

Figure 3.13. The 7451 AND–NOR gate.

Table 3.4. Identification Codes of the TTL Gates

Code	Meaning	Response Time (ns)	Power Consumption (mW)	Maximum Frequency (MHz)
—	Normal	10	10	35
H	High power	6	22	50
L	Low power	33	1	3
S	Schottky	3	19	125
LS	Low-power Schottky	10	2	45

3.5.5. Combinations of Gates

More complex ICs, designed to perform several specific functions, are built up with combinations of these gates. A very common and simple combination is the AND–NOR gate, constituted by two AND gates and one NOR gate. The circuit, truth table, and the 7451 IC, containing two of these gates, are depicted in Figure 3.13.

3.5.6. Gates of the TTL Family

There is a whole family of TTL gates that are distinguished from each other by several characteristics, such as signal propagation time, dissipated power, and maximum working frequency. They are named by one or two letters, which are included in the middle of the basic number of the chip, for example, 70LS04. The meaning of these letters is given in Table 3.4. The appropriate chip is selected in accordance with the requirements of the circuits; for example, for portable instruments low power dissipation can be important and thus a low power "Schottky" may be the best choice.

3.6. LATCHES

3.6.1. Introduction

Most of the currently used digital circuits are synchronous; that is, they work on a step-by-step or clock-controlled basis. When a change in the input conditions is produced, synchronous devices do not generate an output immediately afterward. Instead, they wait until a pulse coming from the system clock arrives. Advantages of the clock-controlled logics are:

Nonconfirmed variations of logic level, such as an accidental momentary transition of voltage, cannot reach the outputs.

Changes can only go on in an ordered way, allowing the use of shift registers and counters.

Each function within a system takes place at a given time.

A latch, flip-flop, or bistable is an element that reaches a logic state (0 or 1) when it receives the appropriate input signal, remembering it after the input has ceased; that is, the output maintains its logic level until an opposite input signal is received. In a gate, input signals must be kept continuously active to obtain a given continuous output, whereas this is not necessary in a flip-flop. Computers use large amounts of flip-flops as memory elements.

A typical flip-flop is schematized in Figure 3.14, where D (data) is the input, Q the output, and $\overline{Q}$ the complementary output, which always gives the logic state that opposes the one given at Q. The clear and preset inputs allow one to reduce Q to logic 0 and 1, respectively, under any conditions. When input D establishes that a change must occur, the clock input enables the change of output. The change is achieved when a clock pulse arrives. D is a synchronous input, since it affects the output only under clock control, whereas the clear and preset inputs are asynchronous, because they can affect the output independently of the clock.

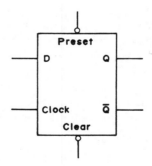

Figure 3.14. A simple latch or flip-flop.

A flip-flop built up with two NAND gates is shown in Figure 3.15. When the switch is in position C, the upper input of B will be grounded and therefore, since the lower input of B is also at low, its output will be at 1. On the other hand, since D is not grounded, the lower input of A is at 1. It can be considered that this input is connected to a voltage divider with two resistances, one of them with a value R and the other one, the floating D, with an infinite value. Since the other input of A is connected to the B output, the A output is at 0.

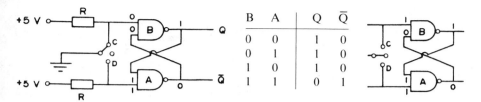

B	A	Q	$\overline{Q}$
0	0	1	0
0	1	1	0
1	0	1	0
1	1	0	1

Figure 3.15. A latch with two NAND gates.

With the switch at an intermediate position (not connected), the inputs are the same, with the exception of the upper input of B, which will now be at 1. However, since the lower B input is at 0 (connected to the A output), the outputs do not change. Obviously, the circuit has a memory, the outputs being reversed only when the connection between D and ground is closed.

3.6.2. Timing Diagrams

The representation of a digital signal must show the changes of states produced as a function of time. These representations are known as timing diagrams. Some conventions usually adopted in these diagrams are:

Time increases from left to right.

The lower states or base lines are considered to be always at the zero level, and the higher states at level 1.

The changes from 0 to 1 and from 1 to 0 occur suddenly under the form of upward, or positive, and downward, or negative, edges, respectively. Since changes of state can take place at a high rate, when a change takes place it is said that it is "strobed."

An upward and a downward edge, together with a clock pulse train (which is simply a square wave), are represented in Figure 3.16. The pro-

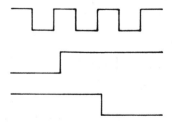

Figure 3.16. A timing diagram.

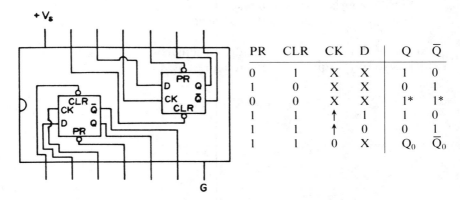

Figure 3.17. Equivalence between circuits using inverters.

pagation delay in a device or circuit is the time required for a digital pulse or signal to pass through it. For a normal or noninverted input, a logic 1 turns a device or circuit on, and a logic 0 turns it off, whereas for an inverted input, a logic 0 activates the device and vice versa. In each case, an upward or downward edge, followed by a negative clock pulse, is required for a change to be strobed at the output. Inverted or reversed inputs and outputs are expressed by a small circle, as shown in Figure. 3.17.

3.6.3. Description of Some Common Latches

The 7474 IC is depicted in Figure 3.18. This IC is provided with clear and preset inputs and contains latches that are strobed with a positive edge. The truth table describes the outputs of each independent latch when different signals enter through the inputs. In the table, Q_0 and $\bar{Q}_0$ represent the previous logic levels and they mean that no change occurs; the prompt means a nonstable configuration that will change when the appropriate clock signal arrives; the arrow pointing upward means strobed by positive edge clock

PR	CLR	CK	D	Q	$\bar{Q}$
0	1	X	X	1	0
1	0	X	X	0	1
0	0	X	X	1*	1*
1	1	↑	1	1	0
1	1	↑	0	0	1
1	1	0	X	Q_0	$\bar{Q}_0$

Figure 3.18. The 7474 IC, showing the two internal latches, and truth table for the latches.

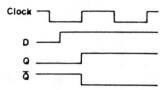

Figure 3.19. Timing diagram of the 7474 latches.

input, and X, an indifferent state, means no changes are produced. When
the preset or the clear inputs are at low, the effects of the clock and the D
inputs are suppressed, leading to ambiguous outputs. The timing diagram
for PRESET = CLEAR = 1 is given in Figure 3.19. If D is at 1, the output Q
changes to 1 when the positive edge of the clock arrives.

Another useful flip-flop circuit, the 7475 IC, is described in Figure 3.20.
The chip contains four latches, two of them strobed through pin 4 and the
other two through pin 13. If both pins are connected to each other, the four
latches can be strobed simultaneously.

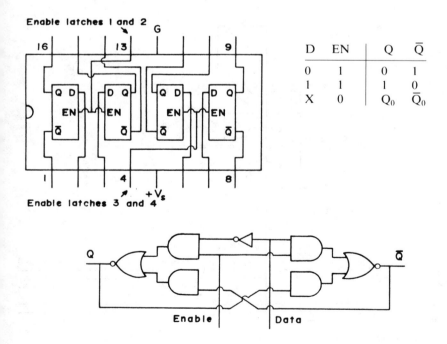

Figure 3.20. The 7475 IC. At the right, the truth table for each latch and internal structure of
the latches, showing the six gates.

The difference between a 7475 latch and a 7408 AND gate is that the former has memory and retains the logic states instantaneously communicated through the inputs, whereas the AND gate has no memory. The differences between the 7474 and 7475 ICs can be deduced from Figures 3.20 and 3.21, and from the timing diagram of Figure 3.22. Unlike the 7474, the 7475 has no preset or clear inputs. As shown in the timing diagram, the 7474 output goes to high, and is maintained at high, when D is at high and the upward clock edge arrives. Similarly, it goes to low, and is maintained at low, when D is at low and the upward clock edge arrives. On the other hand, the 7475 output follows the D logic states when the clock is at high. However, it is locked at high when D is at high and a downward negative clock edge arrives.

The 74100 circuit (Figure 3.23) is a 24-pin latch that contains 8 bistables, each one identical to the 7475. This circuit has no complementary $\overline{Q}$ outputs and the latches are strobed in two 4-latch groups. This 8-latch circuit is very useful to build up 8-bit word microprocessors.

Circuits 74174 and 74175 are flip-flops similar to the 7474, but they contain 6 and 4 latches, respectively. They have clear but no preset, and the clock strobes simultaneously all the internal flip-flops with an upward edge.

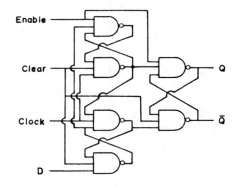

Figure 3.21. Internal structure of the latches of a 7474 IC.

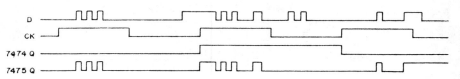

Figure 3.22. Timing diagrams of the outputs of the 7474 and 7475 ICs.

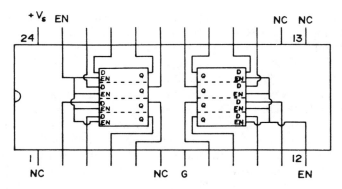

Figure 3.23. The 74100 latch.

A group of flip-flops that are strobed and zeroed simultaneously works as a register. It stores as many bits of information as flip-flops it contains. As we shall see later in this book, the concept of register is very important in relation to microprocessor architecture and functioning.

3.7. ENCODER–DECODERS

Digital codes can be considered as the digital language for storage, handling, and communication of information. Some simple and common digital codes were given previously. In digital electronics, it is often necessary to transform data from one code to another code. For this purpose, code converters are used. In a code converter, data encoded in a given code are fed to the inputs and are received in another code at the outputs.

A very common way of representing binary numbers in decimal form is to use a 7-segment display (see Figure 3.24). The position of the segments

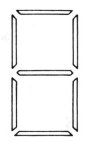

Figure 3.24. The 7-segment display.

permits the representation of the first 10 decimal numbers (from 0 to 9) when the appropriate BCD combination is applied to the decoder driver of the IC display. The same 7-segment displays can also be used to represent hexadecimal numbers, since the digits A to F can also be given by the 7-segment display (b and d as lower case types).

The 7447 IC is a BCD-to-7-segment decoder driver; that is, it is designed to interface BCD signals with 7-segment displays. As shown in Figure 3.25, the 7447 accepts 4-bit BCD data and uses them to determine which decimal digit is intended for display. Pin connections are also shown in the figure. Pin 3 (lamp test) is used for test purposes. It is usually kept at high and turns on all segments when it is at low. In fact, the 7447 is a code converter; however, because the output can be understood directly, it is called a decoder.

The 7442 IC is a BCD decoder that converts 4-bit BCD words into decimal digits. In a 7442, only one of the 10 outputs can be at logic 0, the other 9 outputs being at logic 1. The scheme of the decoder and its truth table are given in Figure 3.26.

Another common decoder is the 74154 IC, which contains four data inputs, two enable inputs, and 16 outputs. The 74154 is a BCD-to-hex decoder that converts a 4-bit word into a 16-bit word that can contain only one logic 0.

The more important features of a code converter are selection time, strobing time, and dissipated power. The selection time is the time elapsed since the application of all the logic states to the inputs, until the specific output has been produced. The strobing time is the time elapsed since the application of the input strobing signal, until the encoder or decoder is actually strobed.

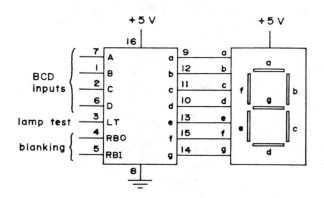

Figure 3.25. The 7447 BCD-to-7-segment display decoder driver.

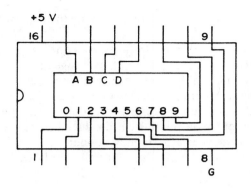

D	C	B	A	0	1	2	3	4	5	6	7	8	9
0	0	0	0	0	1	1	1	1	1	1	1	1	1
0	0	0	1	1	0	1	1	1	1	1	1	1	1
0	0	1	0	1	1	0	1	1	1	1	1	1	1
0	0	1	1	1	1	1	0	1	1	1	1	1	1
0	1	0	0	1	1	1	1	0	1	1	1	1	1
0	1	0	1	1	1	1	1	1	0	1	1	1	1
0	1	1	0	1	1	1	1	1	1	0	1	1	1
0	1	1	1	1	1	1	1	1	1	1	0	1	1
1	0	0	0	1	1	1	1	1	1	1	1	0	1
1	0	0	1	1	1	1	1	1	1	1	1	1	0
1	0	1	0	1	1	1	1	1	1	1	1	1	1
. . . .				. . . .									
1	1	1	1	1	1	1	1	1	1	1	1	1	1

Figure 3.26. The 7442 BCD-to-decimal decoder, scheme and truth table.

3.8. COUNTERS

3.8.1. Introduction

Counting clock pulses is one of the most important tasks in digital electronics. Physical parameters such as resistances, capacitances, voltages, or currents can be digitalized by counting amounts of pulses. In some way, a given number of pulses, counted during a given time period, is always proportional to the measured parameter.

A counter is a device that is able to change the logic state of its outputs by following a given sequence, when the adequate signal is received at the input. A counter is built up with flip-flops and several gates. The outputs of all the flip-flops are always available, allowing the retrieval of the count at any moment.

In a binary counter, all the flip-flops are connected to a single input. Every time a pulse appears at the input, a change of state is produced and tabulated, to generate a multiple n-bit output. The highest possible count is 2^n, where n is the number of flip-flops that the counter contains.

Binary counters can be used to construct devices that perform divisions by n. In an n-divider, an output pulse is produced every n input pulses, and therefore it can be used as a frequency divider.

A decade counter has 10 flip-flops, and the application of pulses to the input produces an output that goes from the decimal 0 to the decimal 9, and then goes back to 0.

Upward counters, such as the 7490 IC, count only upward, increasing the count (e.g., from 0 to 9 and then going back suddenly to 0), whereas up/down counters can increase or decrease the count (e.g., going from 0 to 9 and then back count-by-count from 9 to 0).

In an asynchronous counter, a change in a flip-flop induces a possible change of the following flip-flop and so on. Therefore, a propagation delay is produced. Instead, in synchronous counters, all flip-flops can change their states simultaneously when the clock pulse is received. Another difference between asynchronous and synchronous counters is that the former are strobed by positive clock pulses and the latter by negative ones.

Clear and preset are important functions of counters. They allow the converter to be set at zero (clear) or at a given prefixed value (preset). Thus, for instance, the 7490 IC counter can be set at any value between 0 and 9, whereas the 7493 IC can only be set at 0.

Counters can be chained; that is, the output of a first counter can be connected to the input of a second one, and so on. In this way, four decimal counters mounted in a chain can count until 10,000. The modulus of a counter, n, is the number of different states that the counter repeats; for example, a decade counter has a modulus of 10. Three decade counters mounted in a chain constitute a counter of modulus 1000.

3.8.2.   The 7490 IC Decade Counter

The 7490 IC is a very common decade counter. The connection pins and truth tables are shown in Figure 3.27. The symbol formed by a circle and a triangle, which can be observed at inputs A and B, indicates that these are dynamic inputs, which are strobed by a negative edge.

As shown in the truth table (Figure 3.27, left), each count produces a unity increase in the series of binary numbers. In the timing diagram of Figure 3.28, each dynamic input of the 7490 IC is strobed by a negative edge. Just before

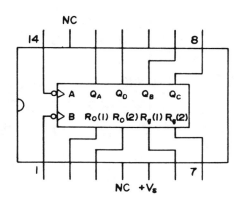

Count	Q_D	Q_C	Q_B	Q_A	R_o (1)	(2)	R_g (1)	(2)	Q_D	Q_C	Q_B	Q_A
0	0	0	0	0	1	1	0	X	0	0	0	0
1	0	0	0	1	1	1	X	0	0	0	0	0
2	0	0	1	0	X	X	1	1	1	0	0	1
3	0	0	1	1	X	0	X	0			counts	
4	0	1	0	0	0	X	0	X			counts	
5	0	1	0	1	0	X	X	0			counts	
6	0	1	1	0	X	0	0	X			counts	
7	0	1	1	1								
8	1	0	0	0								
9	1	0	0	1								

Figure 3.27. The 7490 decade counter. A and B are clock inputs, the R's are response inputs, and the Q's are outputs. The truth table of the left shows the BCD counting order and the truth table of the right, the outputs as a function of the response inputs.

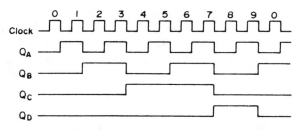

Figure 3.28. Timing diagram of the 7490 IC.

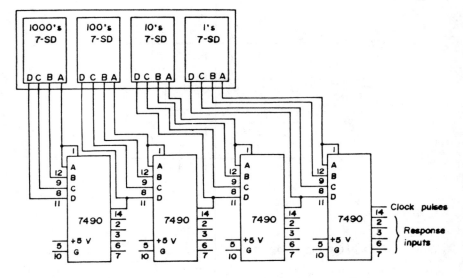

Figure 3.29. A chain formed by four decade counters driving 7-segment displays to generate a 4-digit decimal number.

the negative edge, after the clock high number 5, the values $A=1, B=0, C=1$, and $D=0$ are achieved; that is, $DCBA=0101$, which is the decimal 5.

The downward edge at D produced after the clock high number 9 may be used as an input for another 7490, to control the next counter in a chain of decade counters. It is said that the 7490 IC is a 10-divider counter.

Figure 3.29 shows a chain formed by four decade counters. From left to right, the D outputs of the counters are 10-, 100-, 1000-, and 10,000-dividers. This circuit is able to count from 0000 to 9999. The most significant digit (MSD) is located at the left end and counts thousands; while the least significant digit (LSD) is at the right end and counts units. For the four counters to be cleared simultaneously, pin 2 in each counter is connected to a single source of low and high TTL voltages. To clear the counter, the necessary logic 1 can be provided by means of a manual switch or with a microprocessor.

3.8.3. Other Counters

Counters with a modulus different from 10, such as 2, 5, 6, and 8, are also commonly used. It is interesting to observe that a single flip-flop is a counter of modulus 2. The 7492 IC contains two counters of modulus 6. A simple chronometer or 60-minute clock can be built up by combining one of the 6-dividers of a 7492 IC with a decade counter. Thus, a 60-divider is obtained.

3.9. MONOSTABLE MULTIVIBRATORS

3.9.1. Introduction

A monostable multivibrator (MSMV) is a digital device that provides a single stable voltage or logic state, the other state being nonstable. When a MSMV is strobed, the nonstable state is maintained only along a short period and then the output goes back to the stable state. MSMVs are thus distinguished from clocks, oscillators, or astable elements which do not give any stable state, since they provide the two logic levels in alternate short periods. MSMVs are also distinguished from latches, flip-flops, or bistables, whose outputs can provide both logic levels in a stable form.

MSMVs are analog–digital hybrid circuits, where the digital output is determined by the time constant of an external analog RC circuit. They are used to generate pulses of known lifetime. Common MSMVs are the 555 timing IC and the 74121, 74122, and 74123 ICs. The 555 is used to obtain large pulses, from a few microseconds to several hours. The 7412X series is used to generate short pulses, within the range of 40 ns to 10 μs. Pulse lifetime repeativity is better than 0.5%.

3.9.2. The 74121 Monostable Multivibrator

Pin connections and the truth table for the 74121 IC are given in Figure 3.30. When an input pulse strobes the chip, an output pulse is produced. The lifetime of this pulse depends on the RC time constant. As long as the RC circuit is charging up, the chip cannot be strobed again, thus giving pulses with a constant lifetime, independent of the duration of the input pulses. The chip

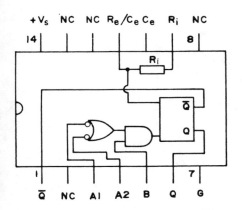

A_1	A_2	B	Q	$\bar{Q}$
0	X	1	0	1
X	0	1	0	1
X	X	0	0	1
1	1	X	0	1
1	↓	1	⊓	⊔
↓	1	1	⊓	⊔
↓	↓	1	⊓	⊔
0	X	↑	⊓	⊔
X	0	↑	⊓	⊔

Figure 3.30. The 74121 monostable multivibrator.

has two inputs that are strobed by downward edges and another one that is activated by upward edges. Two complementary outputs are obtained.

To operate the chip, an external capacitor must be connected between C_{ext} (pin 10) and R_{ext}/C_{ext} (pin 11). On the other hand, the resistor is connected between R_{ext}/C_{ext} and V_s (pin 14). An internal resistor may be used instead, by connecting R_{int} (pin 9) with V_s and leaving R_{ext}/C_{ext} unconnected. If a potentiometer is used instead of a fixed resistor, pulses of regulable lifetime are obtained.

A possible way of mounting the 74121 IC is shown in Figure 3.31. Since A_1 and B (pins 3 and 5) are always at level 1, from the truth table (see Figure 3.30) it is deduced that, when a negative edge is applied to input A_2 (pin 4), a pulse is produced at the outputs. The relationship between R_{ext}, C_{ext}, and pulse

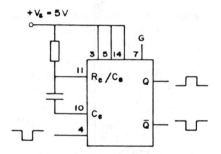

Figure 3.31. A monostable multivibrator circuit.

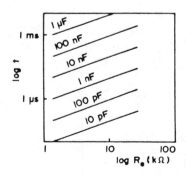

Figure 3.32. The 74121 IC: relationship between pulse lifetime and R_{ext} for several values of C_{ext}.

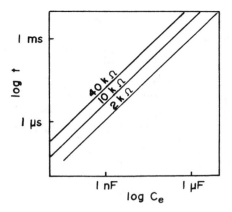

Figure 3.33. The 74121 IC: relationship between pulse lifetime and C_{ext} for several values of R_{ext}.

lifetime can be obtained either from the diagrams given by the manufacturer, which are shown in Figures 3.32 and 3.33, or calculated from

$$t = 0.693\, R_{ext}\, C_{ext}, \tag{3.1}$$

where t is in seconds, R_{ext} is in ohms, and C_{ext} is in farads.

3.9.3. The 74122 Monostable Multivibrator

The 74122 IC is shown in Figure 3.34. This monostable multivibrator is provided with preset and clear. The preset input allows the chip to be strobed again, even before the end of an output pulse. As a result, the pulse is extended in an additional lifetime period. This process can be repeated endlessly, thus producing a stable output as long as strobing signals continue to arrive at a sufficiently high frequency.

Monostables provided with preset are used to detect malfunction, or momentary failures, of synchronous digital circuits, that is, systems controlled by a central clock. Computers and digital instruments are synchronous systems. The RC time constant of the monostable can be adjusted for the output to be at high, as long as the clock feeds one of the inputs with strobing signals. When, for any reason, the clock pulses stop, the output goes back to low, triggering a sequence of events. In this way, the computer or digital instrument can be protected and the memory contents saved before the power vanishes completely.

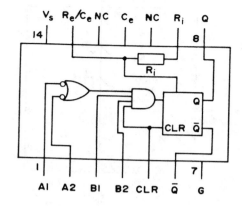

Figure 3.34. The 74122 monostable multivibrator with its truth table.

Clear	A_1	A_2	B_1	B_2	Q	$\overline{Q}$
0	X	X	X	X	0	1
X	1	1	X	X	0	1
X	X	X	0	X	0	1
X	X	X	X	0	0	1
X	0	X	1	1	0	1
1	0	X	↑	1	⎍	⎍
1	0	X	↑	↑	⎍	⎍
1	X	0	1	1	0	1
1	X	0	↑	1	⎍	⎍
1	X	0	1	↑	⎍	⎍
1	1	↓	1	1	⎍	⎍
1	↓	↓	1	1	⎍	⎍
1	↓	1	1	1	⎍	⎍
↑	0	X	1	1	⎍	⎍
↑	X	0	1	1	⎍	⎍

3.9.4. The 555 Timing Circuit

The 555 is a circuit that gives a square wave at the output. As was shown in Chapter 2, the frequency of this wave is established by means of an external RC circuit. Therefore, the 555 is a timer, and oscillators, monostable multivibrators, pulse generators, and other useful devices can be constructed with it.

The 555 has an excellent thermal stability and can be powered with a wide range of voltages (from 4.5 to 16 V). The large output current (up to 200

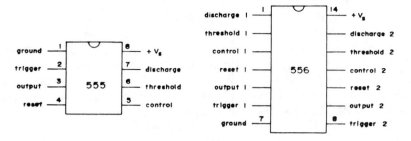

Figure 3.35. The 555 and 556 pin connections.

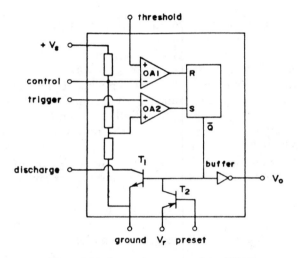

Figure 3.36. Internal structure of the 555 IC.

mA) permits direct control of many loads simultaneously. The 556 chip has two 555 circuits. The pin connections are given in Figure 3.35. The internal structure of the 555 is shown in Figure 3.36. It contains two OAs working as voltage comparators, a bistable with a complementary output $\bar{Q}$, two switching transistors, an inverter buffer for the output, and a voltage divider with three identical resistors.

Since these resistors are equal, the voltages applied to the OA inputs are $V_- = \frac{2}{3}V_s$ and $V_+ = \frac{1}{3}V_s$, for OA1 and OA2, respectively (see Figure 3.36). The 555 has a single supply, the two possible states of the comparators being V_s and 0 V. When the potential at the noninverting input of OA1 becomes higher than the threshold value, $\frac{2}{3}V_s$, the output goes to high. Since this output is connected to the input R of the bistable, a level 1 is also obtained at $\overline{Q}$. Thus, the transistor T_1 becomes saturated and the output goes to level 0.

On the other hand, the voltage at the inverting input of OA2 constitutes another threshold, called the low threshold or trigger. If this voltage falls below $\frac{1}{3}V_s$, the input S of the bistable goes to level 1 and $\overline{Q}$ goes to zero. Transistor T_1 is blocked and the output goes to level 1.

The 555 output can be set to zero by applying a level 0 to the preset. When this occurs, transistor T_2 is saturated, which blocks T_1 and produces the zero output.

3.9.5. Use of the 555 as a Monostable Multivibrator

Use of a 555 as a monostable multivibrator is illustrated in Figure 3.37. Initially, the output of the bistable, $\overline{Q}$, is at level 1 (see Figure 3.36). Therefore, $V_o = 0$ and the LED is off. Simultaneously, discharge transistor T_1 is saturated and the external capacitor, C_1, is not charging. This situation does not change until the START button is pushed. In this moment, S goes to 1, $\overline{Q}$ to 0, and V_o to 1, which is indicated by the lighting LED. Transistor T_1 is blocked and C_1 charges up exponentially through R_3 and P.

When the capacitor charge is $\frac{2}{3}V_s$, the output at R goes to 1, and the bistable goes back to the initial situation, where $\overline{Q} = 1$ and $V_o = 0$. Transistor T_1 goes to saturation, which produces the almost instantaneous discharge of C_1. The time the output has been at high is given by

$$t = 1.1(R_3 + P)C. \tag{3.2}$$

The minimum value of $R_3 + P$ must not be smaller than 100 Ω so that the instantaneous currents through T_1 stay below 200 mA.

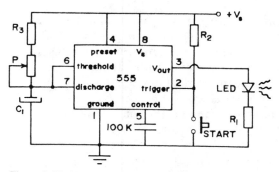

Figure 3.37. Use of a 555 as a monostable multivibrator.

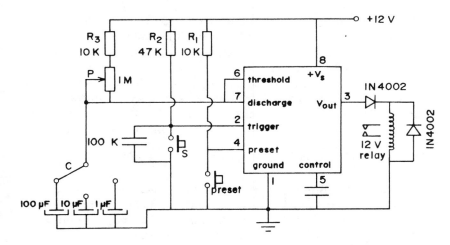

Figure 3.38. A timer circuit with three frequency scales.

The circuit in Figure 3.38 is a practical timer of tripple scale. The three frequency scales are provided by switch C, which can be connected to three different capacitors. Let us see next how the scales can be established.

The lifetimes of the highs or pulses are fixed by a combination of the capacitances at C and the resistances at R_3 and P. The potentiometer P establishes the minimum and maximum values of each scale. When P is at zero, the pulse duration is the smallest of the scale. If we are interested in having a minimum pulse lifetime of 11 ms with the switch connected to the 1 µF capacitor, the value of R_3 must be

$$R_3 = \frac{11\times10^{-3}}{1.1\times10^{-6}} = 10K.$$

On the other hand, if we are interested in a maximum pulse lifetime of 1 s for this scale, P must be

$$P = \frac{1}{1.1\times10^{-6}} - R_3 \approx 1\ M\Omega.$$

The timing ranges for the other capacitors are calculated in an analogous way.

With this circuit, the 555 output can drive a 12 V relay with a current up to 200 mA. The diodes protect the IC from the large relaxation induced currents of the coil. The timing begins when the push-button S (start) is pressed. The 100K capacitor, which bypasses the button contacts, avoids multiple triggering of the circuit during the time the start is pressed. The timing can be interrupted at any moment by means of the preset, or push-button R. When the preset is pressed, the 555 internal transistor T_2 is saturated. This circuit is inadequate for very long timing periods.

3.9.6.   Use of the 555 as an Astable Multivibrator

An astable multivibrator is an oscillator that generates square waves and that can be used as a clock. The word astable indicates that neither of the two logic levels is stable. Use of the 555 as an astable multivibrator is illustrated in Figure 3.39. When the circuit is powered, the capacitor C is discharged, and the voltage at the trigger threshold pin is zero (see also Figure 3.36). In these conditions, R is at low and S at high, and therefore $\bar{Q}$ is at high (V_{out}).

This situation is maintained until the voltage at C reaches $\frac{2}{3}V_s$. Then $\bar{Q}$ goes to low and C discharges through R_B and the discharging transistor T_1. When the voltage at C reaches $\frac{1}{3}V_s$, the capacitor loading cycle is repeated again. As shown below, the output is a symmetrical square wave. The high lifetime is given by

$$T_1 = 0.693(R_A + R_B)C \qquad (3.3)$$

and the low lifetime by

$$T_2 = 0.693R_B C. \qquad (3.4)$$

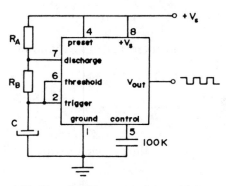

Figure 3.39. Use of a 555 as an astable multivibrator.

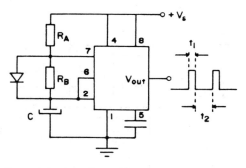

Figure 3.40. Use of a 555 as a pulse generator.

Therefore, if $R_B \gg R_A$, then $T_1 \approx T_2$ and the output is an almost symmetrical square wave with a frequency given by

$$f = \frac{1}{T_1 + T_2} = \frac{1.44}{(R_A + 2R_B)C} \approx \frac{0.72}{R_B C}. \tag{3.5}$$

If a pulse generator is needed, the circuit of Figure 3.40 can be used. This circuit gives $T_1 \ll T_2$. The function of the diode is to make the load of C independent of R_B. Thus,

$$T_1 = 0.693 R_A C. \tag{3.6}$$

On the other hand, the discharge depends on R_B and

$$T_2 = 0.693 R_B C. \tag{3.7}$$

3.10. BUSES

3.10.1. Introduction

Computers and other complex digital circuits are built up with a series of units that exchange data. Thus, in a 16-bit word computer, the central processor unit (CPU), the memory, and several peripherals must be able to send and to receive 16-bit words. It would be awkward to connect each unit to all the other units with 16 separate wires. A better solution is to use a data bus, that is, a single set of 16 wires to which each device is connected. For the bus to work properly, only one device at a time must be allowed to assert data. This is called the talker, and all the other units are only listeners; that

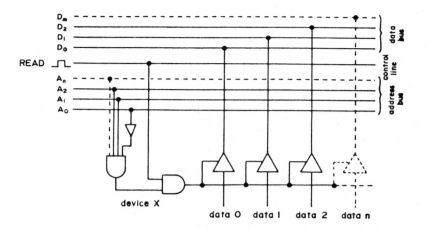

Figure 3.41. A bus, showing the data, address, and control lines. Data enter the data bus through three-state gates.

is, they may only receive data. The agreement of which element may talk is established by means of an additional control line and several address lines, as shown in Figure 3.41.

These different sets of lines are usually referred to as data bus, address bus, and control bus. The data bus transmits data and has as many lines as required by the word length of the computer or instrument, usually 8, 16, or 32. The address bus is used to indicate the origin or the fate of the data. The standard address bus has 16 lines, and therefore it can address a maximum of $2^{16} = 64K$ positions (remember, $1K = 1024$ bits). The control bus is used to synchronize the system.

Obviously, the output of a listener must be disconnected from the bus, as long as it continues to be only a listener. Therefore, gates with active pull-up outputs cannot be used to drive a bus, since the outputs are always connected to the shared data lines. Instead, buses are driven by devices with "open" outputs. There are two types of device: three-state gates and open-collectors. The former are usually preferred for busing, because of their superior performance (speed and noise immunity).

3.10.2. Devices with a Three-State Logic

Three-state devices work on the basis of a three-state logic, which is constituted by the states of high, low, and off. Three-state gates are enabled (put in a state different from off) by means of a separate control line. When

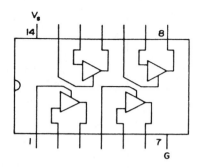

Figure 3.42. The 74126 showing the four internal three-state gates.

enabled, they work as any other standard gate. When they are at off (disabled), their outputs are maintained at a high impedance state.

By using three-state gates, also called buffer drivers, a large number of devices can be connected to a common bus without interference, as long as, at any time, one and only one of them is enabled to output. A typical three-state gate is the 74126 IC (Figure 3.42). The outputs of this IC are enabled only when the control line is at high.

In a bus, each device is wired as the device X. The address of X is a particular combination of logic states of the address lines, A_0–A_n. When the device decodes its address, it knows that it is allowed to assert data onto the data lines D_0–D_m. Then, the device output becomes enabled by the next READ pulse of the clock, which arrives through the control line.

3.11. MULTIPLEXERS AND DEMULTIPLEXERS

These devices can be considered as the electronic substitutes of mechanical multiple switches, as depicted in Figure 3.43. Multiplexers have several inputs and a single output, the reverse being true for demultiplexers.

3.11.1. Analog Multiplexers

In an analog multiplexer, one of several analog inputs is selected by means of an address, that is, a combination of digital address inputs. Only the addressed analog signal is allowed to appear at the output. The number of selectable inputs depends on the number of address inputs available. Thus, for instance, an 8-input multiplexer (Figure 3.43) has a 3-bit ($2^3 = 8$) address

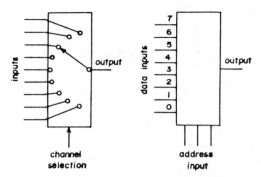

Figure 3.43. An electronic multiplexer (right) and its mechanical equivalent (left).

input. Analog multiplexers are bidirectional; that is, they can also be used as demultiplexers.

The CMOS 4051BE analog multiplexer/demultiplexer is shown in Figure 3.44. It has 8 inputs (D0 to D7) and a single output, which can be reversed to 8 outputs and an input. The selection of the input that is connected to the output depends on the binary configuration at the address lines A0–A2. Thus, the combination 000 will connect D0, whereas 111 (the decimal 7) will connect D7. The internal resistance is 200 Ω, which usually will not produce a significant voltage drop across the device.

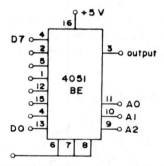

Figure 3.44. The CMOS 4051BE analog multiplexer/demultiplexer.

Figure 3.45 shows an application example of the 4066 IC analog multiplexer. The circuit is a programmable gain amplifier. The three possible analog current paths are selected by the digital address. The normal "on" resistance of the 4066 is 80 Ω, which must be taken into account to implement the external resistors. The values given in the figure have been calculated for gain settings of 10, 100, and 1000.

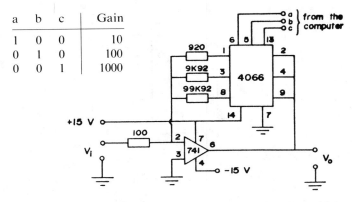

a	b	c	Gain
1	0	0	10
0	1	0	100
0	0	1	1000

Figure 3.45. A programmable gain amplifier built up with a 4066 multiplexer.

An analog multiplexer can also be used to set the gain of an AD524 IC, which is a programmable instrumentation amplifier. The circuit and gain values are given in Figure 3.46. A gain of 1000, 100, or 10 is obtained by connecting pin 3 to pin 11, 12, or 13, respectively. Intermediate gains can be obtained by adding external resistors.

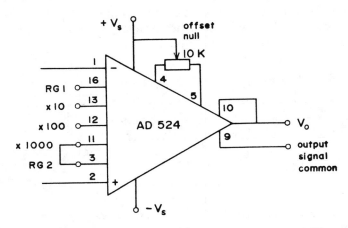

Figure 3.46. The AD524 programmable instrumentation amplifier.

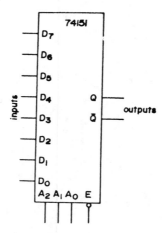

Figure 3.47. The 74151 digital multiplexer.

3.11.2.   Digital Multiplexers

Digital multiplexers provide a logic output that is at the same logic state as the selected input. A given input is selected by means of a particular combination of the logic states of the address inputs. Digital multiplexers are not bidirectional. They are available with 2, 4, 8, and 16 inputs, which correspond to the presence of 1, 2, 3, and 4 address inputs, respectively.

The 74151 IC digital multiplexer is shown in Figure 3.47. It has an enable input (E), and it provides true (Q) and complementary ($\bar{Q}$) outputs. When E is at high the chip is disabled, which means that Q is at low and $\bar{Q}$ is at high, independently of the states of the inputs.

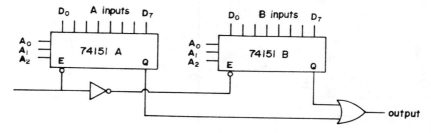

Figure 3.48. Connection of two digital multiplexers to provide 16 inputs and a single common output.

Figure 3.48 shows how two multiplexers can be connected to each other to provide a larger number of selectable inputs. When one of the chips is enabled, the other one is disabled and vice versa, due to the inverter placed between the enable inputs. A single output may be obtained by means of an OR gate connected to the Q outputs.

A digital demultiplexer performs the opposite function: it takes an input and addresses it to one of the diverse possible outputs. The other (not addressed) outputs are either held in the inactive, high-impedance state or at open circuit, depending on the type of demultiplexer.

CHAPTER

4

MICROPROCESSORS

4.1. INTRODUCTION

A microprocessor (µP) is an IC that contains the necessary internal components to perform a series of control, arithmetic, and logic operations. A microprocessor is, by itself, an on-chip IC computer, and today it typifies the most advanced kind of IC available. A µP is a computer central processing unit (CPU) that includes an arithmetic unit, several registers (some of them are on-chip memories), and analog input/output facilities (I/O). All that must be done to construct a microcomputer is to add larger data storage unities and peripherals (console, keyboard, printer, etc.) around a µP.

The design of all kinds of instruments has been revolutionized by the inclusion of µPs, which today can be found as dedicated devices in almost every piece of equipment. A µP-based instrument delivers better performances at lower cost. Also, substitution of mechanical parts and discrete logic chips by µPs simplifies equipment manufacture.

The history of the µP is interesting and illustrative. The µP is neither the result of good planning, commercial foresight, nor the product of an ingenious mind. It appeared gradually and accidentally. Thus, the first µP, the Intel 4004, introduced in 1971, was the result of a contract with a Japanese manufacturer of table calculators. The performance of the 4004 was very limited and unsuitable for general purposes. The Intel 8008, an 8-bit µP, appeared in 1972 as a result of a contract with Display Terminals Corporation (Datapoint) for the development of a single IC capable of controlling a cathode ray tube (CRT). The resulting chip was too slow and Datapoint chose another solution. However, sales of the 8008 IC began to increase and its possibilities were clearly understood. A race for the development of more powerful and rapid µPs began.

Although other manufacturers have developed remarkable µPs, such as the 68000 by Motorola and the Z-80 by Zilog, a good idea of the evolution of the µP can be given by following development of the Intel µPs. After the 8008, the 8080 and the 8085 appeared. Some auxiliary chips, such as the clock, were incorporated into the single µP chip, and thus only the external quartz crystal should be added to an 8085 µP. The introduction of the 16-bit µPs, such as the 8086 and 8088 chips, and, finally, the 32-bit µPs, such as the 80386, represents an enormous expansion of µP possibilities.

The structure and mode of operation of µPs are described in this chapter. Their essential components are described first, introducing the necessary concepts and related terminology.

4.2. HARDWARE, SOFTWARE, AND FIRMWARE

Hardware or hard refers to the physical components of the system, whereas software or soft consists of the programs. The firmware is a special kind of software. The microprograms that regulate the operation of a µP and that are internally recorded in the chip are firmware. The concept of firmware is also applied to any program located in the passive or read-only-memory (ROM) of a computer and that cannot be modified by the user.

A microprogram is a program that resides in the control unit of a µP. Usually its function is to interpret the external instruction index. The external instruction index is a list of instructions which the user can employ to elaborate the programs.

4.3. MEMORY

The memory of a system is used to store data and instructions. There are several kinds of memory:

1. *Internal registers*, usually located inside the main µP chip or CPU. This is the most rapidly accessible memory with typical access time below 100 ns.

2. *Main memory*, constituted by one or more additional chips, with a capacity ranging from 256 bytes to more than 1 Mbyte, and with access times of about 300–600 ns. It is usually called the memory of the microcomputer.

3. *Massive memory*, based on peripheral devices, specially designed to store huge amounts of information on low-cost supports. Popular memory peripherals are the diskettes or floppy disks and the magnetic tapes or cassettes. Optical compact disks based on laser recording and reading have a very high storage capacity, although they can be recorded only once. Micro and minicomputers also use hard disks to store frequently needed information, including programs and data.

Access time is somewhat lower compared to other types of disk and tape.

Memory is logically organized in the logic information units called words. Common memories have word lengths of 4, 8, 12, 16, or 32 bits. The logic organization of a memory is shown in Figure 4.1. The position of a bit in a datum is represented by a decimal digit within the range 0–n, where $n + 1$ is the word length. This notation is used to represent the weight of the bit in binary. Thus, for an 8-bit μP we can have the following 8-bit datum:

$$1 \quad 0 \quad 1 \quad 1 \quad 0 \quad 0 \quad 0 \quad 1$$

$$2^7 \quad 2^6 \quad 2^5 \quad 2^4 \quad 2^3 \quad 2^2 \quad 2^1 \quad 2^0$$

The bit located at the far right position is the least significant bit (LSB) and the bit at the far left is the most significant bit (MSB).

Each individual datum is stored in a particular position of the memory, which is characterized by its address. Addresses are also given in bits. No relationship exists between the number of bits of the data and the number of address bits.

A datum coming through the data bus is individually addressed by the address bus. At the same time, the control bus indicates that it must be written and not read, and the datum is stored in the selected position of the memory. The time needed to write a datum is called the memory cycle time.

A complete instruction can fill several positions of memory. Thus, when the control unit of the μP reads a two-word instruction, it performs two consecutive memory reading operations.

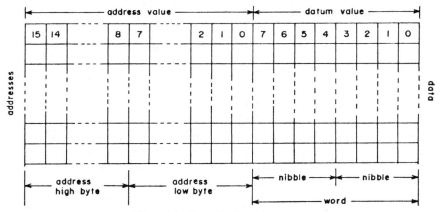

Figure 4.1. Organization of a memory.

4.4.   ARCHITECTURE OF THE INTEL 8085 μP

4.4.1.   Introduction

The Intel 8085 μP is used here to explain the internal structure of the μPs. Reasons to choose this μP are simplicity and the large number of designs based on it. Besides, the 8085 is the predecessor of the 8086 and 8088 μPs, which are used in many present-day personal computers (PCs). The most advanced generation of the family, the 80286 and 80386 μPs, are the base of the PC/AT computers and the IBM personal systems such as the PS/70 and PS/80 models. Some automation examples based on simpler computers, such as the Commodore models VIC-20 and 64, are given later, together with some explanation about the 65XX μP family.

4.4.2.   The 8085 μP Parts

Figure 4.2 shows a block diagram of the 8085 μP. The internal 8-bit data bus and some parts connected to it become evident at first sight: an accumulator, the arithmetic logic unit (ALU), the instruction register, and a register array containing both 8-bit and 16-bit registers.

Instructions can operate on these 8-bit registers individually or in pairs. In the latter case, 8-bit registers are combined to form 16-bit register pairs. Even though the 8085 is basically an 8-bit computer, 16-bit registers are used to address up to $2^{16} = 65536$ memory locations.

The μP also contains a timing and control section to harmonize the activities of the internal data bus and the external control lines in response to the output of the instruction decoder.

The accumulator or register A is the main register of the μP. Arithmetic and logic operations are performed in the ALU with the help of the accumulator and the temporary register. The flag register is essential for all forms of testing and branching operations. The register pair HL performs a "pointing" function for the memory. The memory location whose address is pointed by the HL pair is used in most arithmetic operations. In general, the register pairs (BC, DE, etc.) are used for the 16-bit arithmetics, to handle addresses, and for such operations as counting.

The 8085 μP also contains a program counter that sequentially indicates the instruction and data addresses to be handled, and a stack pointer that indicates the location of the data in a special part of the memory where data and addresses are provisionally stored (e.g., instructions of subroutine returns).

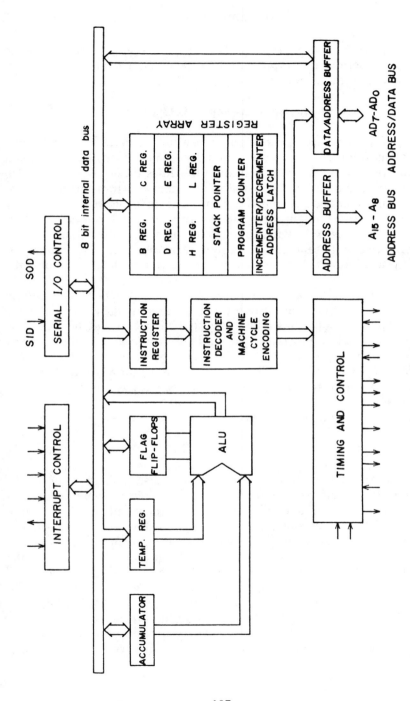

Figure 4.2. Simplified internal structure of the 8085 μP.

107

The interrupt control (upper part of the figure) is a special feature of the 8085 µP; it permits interruption of the system and activation of a control interface for serial communications (SID/SOD).

4.4.3. Bus Structure and ALU Operation Mode

The 8085 µP is packed into a 40-pin chip, which deserves some additional comment. In order to fit the µP into the 40-pin package and to have still enough pins available for the desirable input lines, the designers decided to use the 8-bit data bus pins for a double duty, that is, to transmit data and to support half of the address bus. Since an address is 16 bits wide, the top 8 bits of the address are brought out on their own pins, with the bottom 8 bits sharing the same pins as the data, although multiplexed in time, thus reducing external wiring. To send out an address, the µP uses both, the address bus and the address/data bus, whereas to send or receive data it uses only the address/data bus.

External and internal bus structures differ in a µP. In fact, an important way of classifying µP internal architectures is to count the number of buses that support the communication between the registers and the ALU. The most simple architecture is a single bus. In this case, data coming from any register are transmitted to the ALU through the single bus, which is connected with two ALU inputs. The result of the operation is output through the same bus to be transmitted back to the registers. A single bus must be multiplexed in time.

A typical instruction could be

$$R_0 = R_0 + R_1,$$

which means that the sum of the contents of registers 0 and 1 must be the new content of register 0. The execution of this simple instruction requires a complete sequence of operations, each one of them with a finite duration and which are synchronized along several clock cycles. As depicted in Figure 4.3, first, the content of R_0 is sent to the accumulator; and second, the contents of the accumulator and of R_1 are sent to the ALU. Finally, the result is sent back from the ALU to R_0, where it replaces the old contents, whereas the contents of R_1 remain invariable. The function of the accumulator has been to temporarily memorize the contents of R_0, thus allowing the use of the single data bus for the other data transfer.

Therefore, three transfers are needed to complete an operation and the data bus must be time multiplexed. Obviously, slowness is the drawback of

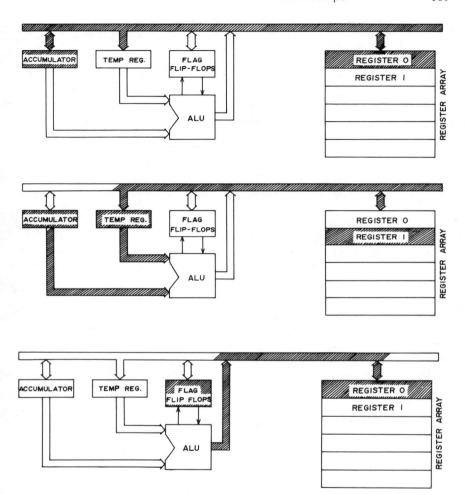

Figure 4.3. An operation sequence when a single internal bus is used. First, the contents of register 0 are sent to the accumulator; second, the contents of register 1 are sent to the ALU and added to the contents of the accumulator; third, the result is sent to register 0.

a single bus system. The main advantage is to require a smaller chip area, which has justified its use until recently.

The ALU also permits execution of displacement and rotation instructions, which are characterized by the movement of the bits inside a byte. In

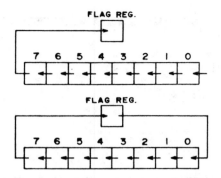

Figure 4.4. Left shifting and rotation operations.

these operations, the bit ejected from the left side of the register, or carriage bit, is stored in a special register of states or flag register, as shown in Figure 4.4.

4.4.4.   The Register of States

The register of states or flag register, which appears over the ALU in Figure 4.2, stores exceptional conditions of the ALU. The content of this register can be verified by means of specific instructions. Flags or bits contained in this register are the carriage bit, and the O, N, H, Z, and P bits.

The carriage bit performs two different functions:

1. Arithmetic carriage, that is, supply of a 9th bit when it is required by an 8-bit arithmetic operation, as shown in the following example:

$$
\begin{array}{r}
1000\ 0000 \\
+\quad 1000\ 0001 \\
\hline
1\ 0000\ 0001
\end{array}
$$

Some instructions, such as "ADD with carriage," automatically add the carriage bit to the result of the following sum.

2. The carriage bit can also be used to detect an overflow when a displacement or a rotation operation is performed.

The O bit or overflow bit reveals the modification of the most significant bit of a word. The N bit or negative bit, is related to the position of the most significant bit of the result, which in some operations represents the sign of a number. The H bit is used only in operations performed in BCD. The Z, or zero bit, is used to verify if the result of an operation is zero and also to perform some logical instructions of comparison. When the result of an operation is zero, the Z bit is put at logic 1.

Finally, the P bit, or parity bit, of the flag register is used to detect if the number of "1" states of a word has the correct parity. If even parity is being used, the parity bit is put at 1 whenever the number of 1 states of the word is not even, thus indicating that a parity error has been produced. This is a very important concept in system communications and will appear later in the section devoted to serial interfacing. Not all the µPs have a parity bit, since, in a computer, the parity detection is frequently performed by an interface chip called UART.

The register of states also contains other 1-bit registers that permit system interruptions. Thus, a logic 1 in an interruption bit enables an interruption, whereas a logic 0 disables or "masks" it.

Most of the instructions executed by a µP have an effect on one, several, or even all these bits, which are used as indicators or "flags."

4.4.5. Internal Registers

Two types of internal register can be distinguished: specialized and nonspecialized. Nonspecialized internal registers are required for the ALU to be able to handle data at high speed. In an 8-bit µP, each one of these registers is constituted by 8 flip-flops that work in parallel and that are connected bidirectionally to the internal bus. Usually, internal registers are used to store 8-bit data, although they can also work by pairs, thus storing 16-bit data or 16-bit addresses.

Specialized internal registers are the two address registers, that is, the program counter (PRC) and the stack pointer (SP), which are double (16-bit) registers. Their most important feature is that they are connected to the external address bus. They can only be loaded through the 8-bit data bus, which means that two transfers are necessary. The bytes are loaded beginning at the least significant one.

All µPs have a PRC that stores the address of the instruction to be executed next. Finally, some µPs also have an internal index register. Indexing is a memory address system that allows access to entire data blocks with a single instruction. The 8085 µP lacks this system.

A stack is a memory with LIFO structure, which means that the last datum stored is the first one to be retrieved. The acronym LIFO stands for "last in, first out." The memory structure of a stack is therefore chronologic; that is, the data are found in the order in which they were chronologically stored. The usual access to a stack is through its upper end. A stack is handled by means of two specific instructions: PUSH and POP. The contents of a register are stored at the upper part of the stack using PUSH, and the contents of this same first upper location are retrieved and transferred to another register using POP. Stacks are necessary to structure interruptions and subroutine levels.

There are two types of stack memory: hard and soft. A set of internal registers in the CPU is required to implement a hardware stack. On the other hand, a part of the RAM memory, which is external to the CPU, may be used to implement a software stack. A memory position is selected, and its address is stored in the stack pointer. This register manages the stack automatically, increasing its count each time a stack input operation (PUSH) is performed, and decreasing it with the stack output operations (POP). Actually, the stack pointer always points to the first empty memory position which is located immediately above the last occupied position, and thus when a PUSH instruction arrives, it is immediately executed.

In comparison, a hardware stack has a much higher access rate, but the number of available registers is usually small. In a software stack, almost all the RAM memory can be reserved for stacking.

4.4.6.   Instruction Register, Decoder and Control Units

Each μP instruction is executed in a three-step sequence: search, decoding, and execution. First, the instruction is searched for in the memory and transferred to an instruction register (IR) of the μP, where it is decoded. Next, a sequence of signals for the appropriate execution of the instruction is generated by the central control unit (timing and control, Figure 4.2).

An instruction can be constituted by several bytes. The control unit must access to the memory several times to read the successive bytes of a long instruction. The first word in an instruction always contains an operation code, or "opcode," which indicates the number of times the control unit must access the memory looking for other bytes.

The program counter controls the automatic linking of the instructions during the execution of a program. Each time an instruction is executed, the program counter increases its count in 1. In fact, it points always to the instruction next to that being executed. A control transfer, or control skip to a given memory location (e.g., a GOTO or GOSUB in BASIC), can deliberately be specified by loading the program counter with the adequate address.

The whole system is harmonized by the control unit (CU). It generates synchronization signals for the ALU, the memory, and the I/O circuits; it decodifies and executes the instructions and communicates with the external world through the control bus I/O lines. A CU contains an internal program or firmware, which sequences its functions. Elaboration and introduction of this program into the CU is called microprogramming.

μP operations are synchronized; that is, they are performed along a given number of machine cycles. For instance, in an 8085 μP, a search operation, which implies transfer of an instruction from the memory to the decodifier, takes one machine cycle. The complete execution of an instruction takes from one to five machine cycles, depending on how long the instruction is (1, 2, or 3 word length) and how many additional accesses to the memory it requires.

Each machine cycle is performed internally, by means of a microinstruction sequence. Each step of this sequence corresponds to the execution of a microinstruction and is called an internal state. Each machine cycle requires from three to five internal states. Microinstruction sequences are synchronized by the clock of the system in such a way that each internal state corresponds to the period between two successive clock pulses.

4.5. μP INSTRUCTIONS

The list of internal instructions of a μP constitutes one of its essential features. The instructions of the 8085 μP are given in Table 4.1.

One-word instructions are, in principle, the most rapidly executed. An example would be MOV A,C, which means to transfer the contents of register C to the accumulator. The instructions are internally executed in binary code, however they are represented in a symbolic or mnemotechnique way, which is called assembler language. Programming in assembler is easier and safer than in binary. For instance, MOV A,C is equivalent to the hex 79 and to the binary 01 111 001. The first 01 stands for MOV and the following 111 and 001 stand for A and C.

In a one-word instruction, such as this one, there is no need for indicating any memory address. On the contrary, instruction MVI A,18H has two words, the first one (MVI A) is the opcode, and the second one (18H), the hex number 18, is the address. The instruction says that this hex number must be moved (loaded) to the location immediate to the accumulator. Hex and binary equivalents are:

Mnemonic	Hex	Machine Language
MVI A	3E	0011 1110
18H	18	0001 1000

Finally, CALL DELAY is an example of a three-word instruction. It performs a call to the delay subroutine, which is located at a given address of

Table 4.1. Summary of Instructions of the 8085 μP: Mnemonics and Hex Addresses

Data transfer group	Logics and Arithmetics		Branch	I/O and Control
Move[a]	Add[b]	Decrement[c]	Jump	Stack Operations
MOV A,A 7F	ADD A 87	DCR A 3D	JMP ad C3	PUSH B C5
MOV A,B 78	ADD B 80	DCR B 05	JNZ ad C2	PUSH D D5
. . . .	. . . .	. . . .	JZ ad CA	PUSH H E5
MOV A,M 7E	ADD M 86	DCR M 35	JNC ad D2	PUSH PSW F5
MOV B,A 47	ADC A 8F	DCX B 0B	JC ad DA	POP B C1
MOV B,B 40	ADC B 88	DCX D 1B	JPO ad E2	POP D D1
. . . .	. . . .	DCX H 2B	JPE ad EA	POP H E1
MOV M,L 75	ADC M 8E	DCX SP 3B	JP ad F2	POP PSW[b] F1
			JM ad FA	
XCHG EB	Substract[b]	Specials	PCHL E9	XTHL E3
				SPHL F9
Move	SUB A 97	DAA[b] 27	Call	
Immediate[d]	SUB B 90	CMA 2F		Input/Output
	. . . .	STC[e] 37	CALL ad CD	
MVI A,b 3E	SUB M 96	CMC[e] 3F	CNZ ad C4	OUT b D3
MVI B,b 06	SBB A 9F		CZ ad CC	IN b DB
. . . .	SBB B 98	Logical[b]	CNC ad D4	
MVI M,b 36		ANA A A7	CC ad DC	Control
	SBB M 9E	. . . .	CPO ad E4	DI F3
Load		ANA M A6	CPE ad EC	EI FB
Immediate	Double Add[e]	XRA A AF	CP ad F4	NOP 00
LXI B,db 01	DAD B 09	. . . .	CM ad FC	HLT 76
LXI D,db 11	DAD D 19	XRA M AE	Return	New
LXI H,db 21	DAD H 29	ORA A B7		Instructions
LXI SP,db 31	DAD SP 39	. . . .	RET C9	RIM 20
Load/Store	Increment[c]	ORA M B6	RNZ C0	SIM 30
		CMP A BF	RZ C8	
LDAX B 0A	INR A 3C	. . . .	RNC D0	
LDAX D 1A	INR B 04	CMP M BE	RC D8	
LHLD ad 2A	. . . .	Arithmetic and	RPO E0	
LDA ad 3A	INR M 34	Logical	RPE E8	
STAX B 02	INX B 03	Immediate	RP F0	
STAX D 12	INX D 13		RM F8	
SHLD ad 22	INX H 23	ADI b C6	Restart	
STA ad 32	INX SP 33	ACI b CE		
		SUI b D6	RST 0 C7	
Rotate[e]		SBI b DE	RST 1 CF	
RLC 07		ANI b E6	. . . .	
RRC 0F		XRI b EE	RST 7 FF	
RAL 17		ORI b F6		
RAR 1F		CPI b FE		

[a] Dotted lines = the series A, B, C, . . . , M continues.
[b] All flags affected.
[c] All flags except CARRY affected.
[d] b = 8-bit data, db = 16-bit data, ad = 16-bit address.
[e] Only CARRY affected.

the memory. This memory address is 05F1 in hex and has two words. Equivalents are:

Mnemonic	Hex	Machine Language	Meaning
CALL	CD	1100 1101	Opcode
DELAY	F1	1111 0001	Low address byte
—	05	0000 0101	High address byte

This example illustrates how much simpler and safer it is to use a hex instruction than its equivalent in binary for programming purposes.

However, when data and instructions are introduced in hex, a decodifier or interpreter, which translates them to binary, is required. It is still more convenient to introduce the instructions in mnemonic (assembler). However, in this case, a whole assembler–binary translation program is required.

To write the instructions in a code or language closer to the human language implies increasing the complexity of the system, which must translate them to the μP language, which is always constituted by the same binary instructions. A language that approaches the human way of speaking is called a high-level language. The execution of a program written in such a language may be performed by means of an interpreter, that is, a program that translates each instruction before it is executed. However, execution with an interpreter is very slow. A program can be executed much more rapidly when it has previously been compiled, that is, translated to binary. A compiler is a program designed to translate programs from a given high-level language to machine language. Examples are the FORTRAN and BASIC compilers.

4.6. μP-BASED SYSTEMS

The μP is the heart of a microcomputer, but a μP cannot work by itself. Other compatible components are required to perform all the necessary functions. For this purpose, each μP has its own family of ICs, capable of being connected directly to its buses. The three essential components of a system are:

1. The μP itself, which constitutes the CPU, and some additional support components, such as the quartz crystal.
2. The memory, which can be passive or read-only-memory (ROM), and active or random access memory (RAM).
3. The I/O interface circuits, such as the universal asynchronous receptors–transmitters (UARTs) and the programmable adapters,

that is, the programmable interface adapter (PIA), the versatile interface adapter (VIA), and the complex interface adapter (CIA).

Other necessary supporting circuits are latches and buffers. Latches are used when information must be preserved or retained for some time. Buffers (excitators) are required to make up for the limited power of the ICs. The μP bus outputs can support only a single TTL load; thus, intermediate buffers are used to allow the connection of more than one circuit to the buses.

4.7. THE RAM MEMORY

Active or random access memory (RAM) permits one to read or to modify its contents at any moment. The RAM memory is used to store the user's data and programs that are being executed. Its main disadvantage is volatility, which means that its contents disappear when the power supply is cut off. A nonvolatile RAM appeared a few years ago, but it is not popular.

Reading or writing in a memory IC requires previous selection of the IC by means of a chip selection signal (CS) or chip enable signal (CES). This signal, via the control bus, selects one of the memory ICs which are connected to the system buses.

The next step is to give the selected word address inside that memory. Finally, in a reading operation, after the memory access time has elapsed, the datum or instruction searched for appears at the data outputs of the chip. The datum is transferred to the μP through the data bus.

4.8. THE ROM MEMORIES

Once programmed, a ROM memory can only be read. ROM memories are used for the storage of programs and are intrinsically nonvolatile memories. In addition to the simple ROMs, other types of nonvolatile memory can be distinguished: programmable ROM or PROM, erasable-programmable ROM or EPROM, and electrically alterable ROM or EAROM.

ROM memories can be programmed only by the manufacturer: its programming is hardware based and cannot be reprogrammed. Obviously, a relatively high number of identical units must be manufactured to be profitable.

PROM memories can be programmed by the user with the appropriate recording devices. To record a PROM and to set it ready for use take a few minutes; however, once programmed they cannot be programmed again. PROM features are similar to ROM's and frequently both are pin-compatible. This allows the manufacturing of **PROM** prototypes to check ROM programs. When a program is ready for use, the final ROMs are mass manufactured.

EPROM memories are also user-programmable and, in addition, they are reusable. The common types are erased by simple exposure to ultraviolet radiation for several minutes. Exposure is facilitated by the upper quartz window. On the other hand, a special device must be used to record a new program. EPROMs are expensive chips and are noncompatible with ROMs.

Finally, both operations—reading and writing a program electrically— are possible with EAROMs. However, the writing operation is very slow in comparison with the reading operation (on the order of milliseconds for the former and microseconds for the latter). Due to the writing slowness and high prices they are seldom used.

4.9. I/O TECHNIQUES

4.9.1. Polling

Usually, the connection of an I/O device requires an interfacing circuit provided with its appropriate controller and management system. The interface can be built up with several gates and registers, although several general purpose IC interfaces may also be used. The three most common I/O interface management techniques are polling or programmed I/O, interruption, and direct memory access (DMA).

In the polling technique, the I/O devices are connected to the address and data buses of the system. Depending on the type of µP system, the I/O devices must also be connected with some of the control bus lines. In this management technique, the device requiring service is found by successively inquiring at each device. The inquiries are made by the µP, and each device gives back a negative or positive answer.

This inquiring process is called handshaking. Before sending information from one device to another, a state bit of the listening device is checked to find out if the device is ready to receive the information. Also, before reading, the state bit of the talker is checked to find out if the register that must provide the word is actually loaded.

The polling technique is very simple, and special lines are not needed. Furthermore, polling is a synchronous operation, where each device controls the state of the others, thus making communication errors unlikely. The disadvantage of the technique is that it requires a lot of attention by the µP, which in some cases cannot be allowed.

4.9.2. Interruption

When polling is too slow or when the μP wastes too much time on it, the interruption technique may be used. In this technique, the I/O devices take the initiative of asking the μP for service. When the μP detects the interruption application, it performs the appropriate management to give the required service.

The interruption application can be accepted or rejected depending on the state of a qualification bit in the flag register. When the interruption is enabled, the μP must find out which device is applying for service. Since several devices can be asking for service simultaneously, a priority system must also be established.

When an interruption has been accepted, and the μP already knows which device will talk, the μP must interrupt the execution of the program and skip to the appropriate interruption treatment routine. It is said that the interruption is vectorized when the skip address is enabled at the same time the application for interruption is made. When the service is finished, the execution of the program continues at the same point where it was interrupted.

The main advantage of the interruption technique is the small response time. Interruptions are used mainly in systems that operate in real time, where it is very important to achieve a very small response time. On the other hand, an additional circuit and several operations that allow continuation of execution of the interrupted program are needed.

4.9.3. Direct Memory Access

The interruption technique can still not be fast enough in some applications. When this is the case, the direct memory access (DMA) technique may be used. With this technique, the time spent in detecting who is the applicant, in the storage of the necessary data to continue execution, and in skipping to a subroutine is saved.

A DMA controller, that is, a special processor designed to transfer blocks of information, is used. This circuit performs the entire transfer process directly, at hardware speed. The I/O devices, instead of disturbing the μP, send their applications for talking directly to the DMA controller, which interrupts the μP activity with an instruction (HOLD) and takes charge of the system. Then, it performs quick word or data block transfers directly from a device to the memory, bypassing the μP.

4.10. I/O DEVICES

4.10.1. The UART

The universal asynchronous receptor–transmitter (UART) is a device used to convert a serial input into a parallel output, or a parallel input into a serial output. The scheme of a serial-to-parallel conversion is shown in Figure 4.5. The signal arrives as a series of high and low state bits (a bit train), which are read synchronously with the clock pulses. The output provides the same information in the form of bytes, that is, 8-bit packages.

The synchronization of the readings means that, for each half clock cycle, an input bit is read, that is, its state is detected. Each word has a start bit and one or several stop bits generated by the UART of the talker device. These bits allow the UART of the listener device to detect automatically the beginning and the end of a word.

When a device works as a talker, the UART transforms parallel bits, that is, bytes, to transmit serial bits. When this occurs, the UART inserts the necessary start and stop bits. These additional bits are filtered out in the receiving UART.

Frequently, the parity system for detection of transmission errors (see Section 4.4.4) is used. The UART can work with even or odd parity, or without parity. UARTs transmit from 5- to 8-bit length words, with its additional start, stop, and parity bits. The latter may not be used.

The main use of UARTs is as I/O devices for serial communication between teletypes, printers, and modems.

4.10.2. Introduction to Parallel I/O Interfaces

Parallel I/O interfaces are known by several names depending on the manufacturer, for example, parallel or programmable I/O (PIO) and

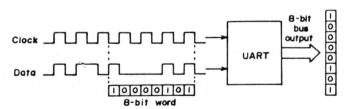

Figure 4.5. Scheme of a serial-to-parallel conversion.

peripheral interface adapter (PIA). They are programmable devices designed to handle parallel bits.

To connect an I/O interface to the data bus it is usually necessary to insert two latches, one for input and the other for output. The input latch retains a word until the μP is ready to receive it. In addition, it separates the coming signals from the bus signals. The output latch is necessary to retain the output datum until the external device is ready to make use of it.

A general purpose parallel I/O interface must have at least one register for input and another for output, status registers, and a logic interrupting circuit. Eight lines are usually not enough with most I/O applications. Many typical applications require at least 16 or 24 I/O lines. For this reason, general purpose interfaces have two or three channels or ports, which can be used for input as well as for output. The interface must be capable of multiplexing these external ports with the data bus of the μP. Each port has its own register, with the proper status information. A scheme of an interface of this type with two ports, A and B, is given in Figure 4.6.

The functions of the control logics of each port of an I/O interface can be programmed. Thus, the user can choose the lines that will be used to

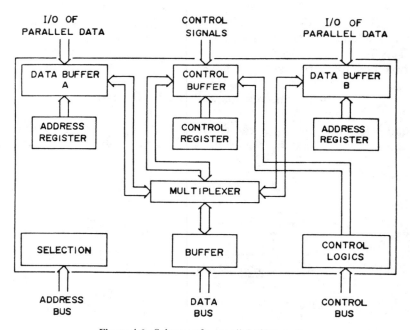

Figure 4.6. Scheme of a parallel I/O interface.

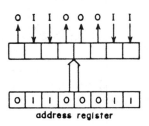

Figure 4.7. Definition of the lines for input or output operation in the 8355 ROM.

dialogue with a device, if the lines will be used to talk or to listen, and, frequently, the function of each individual line. The user can also decide if a port must generate interruptions and pulses or operate with positive or negative logics.

4.10.3. The Intel 8355 ROM with Programmable I/O Facilities

The Intel 8355 is a ROM provided with programmable I/O facilities. Data are sent to, or received from, the I/O lines through data buffers. Each line can be defined individually to send or to receive data. A logic 0 defines the line for input and a logic 1 for output, as schematized in Figure 4.7. The definition of the lines of each port is stored in its own address register. Line programming must be done before transferring the data. The necessary instructions are given in Table 4.2.

4.10.4. The Motorola 6522 Versatile Interface Adapter

The Motorola 6522 IC is called a versatile interface adapter or VIA. This is an interface that can be adapted to a wide variety of situations. It is

Table 4.2. Instructions for Line Programming of the Intel 8355 ROM

Instruction	Function
MVI A,FF	Puts a byte with all its bits in logic 1 in the accumulator
OUT 03	Transfers the contents of the accumulator to the address register of port A; if MVI A,FF has been used first, now all the lines are defined for input
MOV A,C	Transfers the datum stored in register C to the accumulator
OUT 01	Sends the datum through port B

equipped with 16 internal 8-bit wide registers, which are functionally broken down into five specific operations: I/O, timing, shifting, function control, and interrupt control.

Hardware interfacing to the outside world is accomplished by means of two 8-bit I/O ports—port A and port B—and four status/control lines—CA1, CA2, CB1, and CB2—associated with ports A and B, respectively. Both ports are bidirectional and represent an input or an output, depending on the corresponding bit values (0 = input, 1 = output), which are loaded into the data direction registers, DDRA and DDRB. Initially, after a RESET or power-up, all bit values are zero.

Six registers of the 6522 VIA are concerned with timing. The VIA has two internal timers, which can be used as inputs for pulse counters or outputs for pulse generators. Conceptually, the timers may be thought of as counters, each one equipped with a 16-bit register. The hex number placed into the register is decreased by one for each successive clock pulse.

Conversion of serial data to parallel data and vice versa is accomplished by the VIA shift register. The shift register can operate in eight modes, such as enable input serial to parallel and enable output parallel to serial. Selection is made by using the appropriate bit values into an auxiliary control register.

A function control is accomplished by two registers. One is concerned with where signals are going to or coming from, while the other decides how signals are sent or received. These registers are:

1. The auxiliary control register, which provides control over the timers and shift register, and enables/disables data latching on both ports A and B.

2. The peripheral control register, which specifies how the control lines should operate in the input and output modes.

4.10.5. The Complex Interface Adapters (CIAs)

In addition to these classical interfacing circuits, many hybrid ICs can be found on the market. These perform several functions simultaneously, such as memory, PIA, and UART. Other IC interfaces are specially designed for the control of peripherals (keyboards, printers, disks, CRTs, etc.). An example of these is the Intel 8279 for keyboard control.

4.11.   THE INTEL SDK-85 KIT

The Intel SDK-85 kit provides a very simple configuration of a complete microcomputer. A kit such as the SDK-85 is useful, not only to acquire prac-

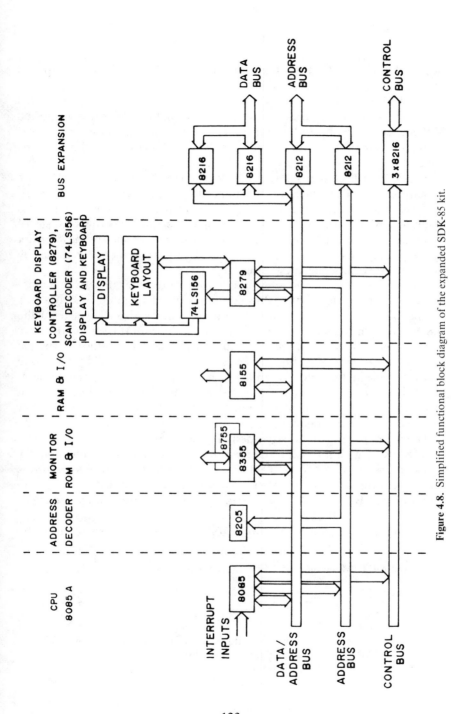

Figure 4.8. Simplified functional block diagram of the expanded SDK-85 kit.

123

tice in the use of µPs and other associated ICs, but also to develop and implement laboratory applications.

A simplified functional block diagram of the SDK-85 in its expanded version is shown in Figure 4.8. All the elements of the system are connected to the three external standard buses, that is, the 8-line data bus, the 16-line address bus (8 of these are shared with the data bus lines), and the 15-line control bus.

The 8085 CPU contains the µP, the clock (connected to an external quartz crystal), and serial I/O interfaces for connection with a teletype (TTY). The µP also has a standard type and four vectored interrupt inputs whose mnemonics and functions are given in Table 4.3. A vectored interruption permits transfer of the execution control to a subroutine whose address is indicated by the interrupt vector. The crystal input to the clock is 6.144 MHz, but it is internally divided by two, the basic clock frequency being 3.072 MHz.

The 8205 IC is a one-out-of-8 decoder, which decodes the 8085 memory address bits to provide chip enables for the 8155, 8355, and the 8279 ICs.

The SDK-85 has the capability of communicating with a teletype using the 8085 serial input and serial output data lines (SID and SOD) to send and receive the serial bit strings that encode data characters. To send data to the teletype, the 8085 must toggle the SOD line in a set/reset fashion, controlled by a software timing routine in the SDK-85 monitor. To receive data, the changes in the level of the SID pin are monitored and timed. Again, a monitor routine is called upon to do the job. Both subroutines communicate at a data rate of 110 baud (bits per second), the standard rate for teletype writers.

The memory map of the expanded SDK-85 kit is given in Figure 4.9. The basic SDK-85 with no additional memory I/O chips provides the memory blocks marked **MONITOR ROM** and **BASIC RAM**. The user's programs must be confined to a subset of the space available in the BASIC RAM, the remainder of BASIC RAM being required for monitor storage locations (monitor reserved RAM).

When an expansion 8155 chip is added in the space provided on the

Table 4.3. Mnemonics and Functions of the 8085 On-Chip Interrupts

Input	Function
RST 5.5	Dedicated to the 8279 IC
RST 6.5	Available user interrupt
RST 7.5	Vectorized interrupt push-button register
TRAP	8055 timer interrupt
INTR	Available user interrupt

**MEMORY
ADDRESS**

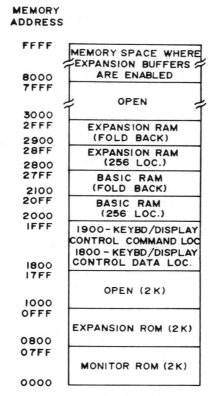

Figure 4.9. Memory map of the expanded SDK-85 kit.

SDK-85 board, the RAM locations shown in Figure 4.9 as EXPANSION RAM are made available. The monitor reserves no space in the expansion RAM; thus all its 256 locations are available for programming.

Any of the areas marked "OPEN" in Figure 4.9 are free for expansion, thus allowing incorporation of extra memory chips. The 8205 address decoder has three uncommitted chip select lines to allow addition of three 2048-byte memory blocks without additional decoding circuitry.

Continuing with the components of Figure 4.8, the 8355 and 8755 are two specially designed chips for compatibility with the 8085 system. The 8355 contains 2048 bytes of mask programmed read only memory (ROM) and 16 I/O lines. The 8755 has an identical function and pinout to the 8355 but

Table 4.4. SDK-85 I/O Port Map

Port	Function
00	Monitor ROM port A
01	Monitor ROM port B
02	Monitor ROM port A data direction register
03	Monitor ROM port B data direction register
08	Expansion ROM port A
09	Expansion ROM port B
0A	Expansion ROM port A data direction register
0B	Expansion ROM port B data direction register
20	Basic RAM command status register
21	Basic RAM port A
22	Basic RAM port B
23	Basic RAM port C
24	Basic RAM low-order byte of timer count
25	Basic RAM high-order byte of timer count
28	Expansion RAM command/status register
29	Expansion RAM port A
2A	Expansion RAM port B
2B	Expansion RAM port C
2C	Expansion RAM low-order byte of timer count
2D	Expansion RAM high-order byte of timer count

contains ultraviolet erasable and reprogrammable read only memory (EPROM) instead of the ROM.

The 8155 contains 256 bytes of RAM memory, 22 programmable I/O lines, and a 14-bit timer counter. One 8155 is included with the SDK-85 kit, and space for another is provided on the circuit board. The RAM memory is available for storage of user's programs as well as for temporary storage of information needed by the system programs. The 8155's timer is used by the SDK-85 monitor's single step routine to interrupt the processor following the execution of each instruction.

The on-board I/O ports of the 8155 and 8355/8755 are accessed using the IN and OUT instructions of the CPU, each individual port being referenced as a unique 8-bit address. Table 4.4 contains all port addresses for the expanded SDK-85 kit.

The 8279 is a keyboard/display controller chip that handles the interface between the 8085 and the keypad and LED displays on the SDK-85 board. The 8279 refreshes the display from an internal memory while scanning the keyboard to detect keyboard inputs.

4.12. MONITOR ROUTINES

A monitor routine is a software program usually located in a ROM. It supervises the operations of a µP and allows humans to communicate with it (usually in hex code). Since it resides in a ROM it cannot be altered. It performs two important functions:

1. Initialization, that is, loading memory locations with certain preset values before execution of certain operations.
2. Linking, that is, establishing dialogue between programs and humans, programs and I/O devices, humans and I/O devices, and between programs.

Table 4.5. A Subroutine of the Monitor Program of the SDK-85 Kit

Function: RDKBD = READ KEYBOARD
Inputs: none
Outputs: A CHARACTER READ FROM KEYBOARD
Calls: nothing
Destroys: the contents of the registers A, H, L, F/F's
Description: RDKBD determines whether or not there is a character in the
input buffer. If not, the function enables interrupts and loops
until the input interrupt routine stores a character in the buffer.
When the buffer contains a character, the function flags the
buffer as empty and returns the character as output.

Loop	Monitor Address	Opcode	Mnemonic	Description
RDKBD	02E7	21FE20	LXI H,IBUFF	Get input buffer address
	02EA	7E	MOV A,M	Get buffer contents
	02EB	B7	ORA A	Is a character available?
	02EC	F2F302	JP RDK10	Yes: exit from loop
	02EF	FB	EI	No: ready for character from keyboard
	02F0	C3E702	JMP RDKBD	
RDK10	02F3	3680	MVI M,EMPTY	Set buffer empty flag
	02F5	F3	DI	Returns with interrupts disabled
	02F6	C9	RET	

Table 4.6. Function of Several Subroutines of the SDK-85 Monitor Program

Calling Address	Mnemonic	Description
0363	UPDAD	Update address field of the display. The contents of the D–E register pair are displayed in the address field of the display.
036E	UPDDT	Update data field with the content of register A (accumulator) in the data field of the display.
02E7	RDKBD	Read Keyboard. This routine waits until a character is entered on the hex keypad and upon return places the value of the character in register A. For RDKBK to work correctly, RST 5.5 must first be unmasked using the SIM instruction.
05F1	DELAY	Time delay. This routine takes the 16-bit contents of register pair D–E and counts down to zero, then returns to the calling program.
02B7	OUTPUT	Output characters to display. The routine sends characters to the display with the parameters set up by registers A, B, H, and L: Register A=0 uses address field Register A=1 uses data field Register B=0 decimal point off Register B=1 decimal point on at right edge of field Register HL starting address of characters to be sent

Usually the monitor program consists of many subroutines that perform specific functions (e.g., scanning a keyboard to detect the depression of a key). Many of these routines can be externally accessed and executed by simply loading the program counter with the starting address of the routine and beginning execution. Once loaded, program control is transferred to that particular subroutine, and a powerful software tool becomes available to the user. Many of these subroutines can be used repeatedly without ever having to code them in the user's program.

Quite often these routines will use the various registers of the CPU for storage of variables. If the same registers are used in the user's program, they must be either reset after exiting the subroutine or saved prior to entering the subroutine. Otherwise, nonsense will prevail.

As an example, a keyboard reading subroutine performed by the SDK-85 monitor program, which resides in the basic ROM, is given in Table. 4.5. In Table 4.6 the function of several useful subroutines of the SDK-85 monitor program are described.

Table 4.7. SDK-85: A Program for Sending Data

Line	Hex	Mnemonics	Description
2000	31C220	LXI SP,20C2H	Initialize stack pointer
2003	3E03	MVI A,03	Put 8155 command in register A
2005	D320	OUT 20H	Program the 8155 CSR
2007	3EFF	MVI A,FF	Put 8355 DDR value in register A
2009	D302	OUT 02	Program port A DDR
200B	D303	OUT 03	Program port B DDR
200D	03	LOOP: INXB	Increment 16-bit count
200E	79	MOV A,C	
200F	D321	OUT 21	Send low byte of count
2011	D300	OUT 0	To 8155 port A and to 8355 port A
2013	78	MOV A,B	
2014	D322	OUT 22	Send high byte of count to 8155 port B
2016	D301	OUT 01	Send high byte of count to 8355 port B
2018	C30D20	JMP LOOP	Loop back

The monitor program of the SDK-85 resides in 2K bytes of the ROM memory between hex locations 0000 and 07FF. Either the keyboard and display or the teletype writer may be chosen as the console device, but not the two peripherals simultaneously.

As explained in Section 4.4.5, the 8085 makes use of a 16-bit internal register called the stack pointer to point to an area of memory called the stack, which is used for saving many things such as memory addresses for returns from subroutines. It is always important to define the stack pointer at the beginning of the program to avoid storing data in the wrong place. Locations 20C2 through 20D0 in RAM are reserved by the monitor for jump instructions when all interrupts are used. Thus, the stack point must initially be at 20C2 by use of the program instruction LX1 SP, 20C2H (31 C2 20 in hex), which means to immediately load the stack pointer with the address 20C2 (in hex). A program that shows how data can be sent through the ports of the SDK-85 is given in Table 4.7.

TRANSDUCTION, SIGNAL CONDITIONING, DATA ACQUISITION, AND CONTROL

5.1. INTRODUCTION

The measurement of a physical or chemical property is conveniently accomplished through a series of several steps: Conversion of the effect into an electrical analog signal or *transduction, conditioning* of this signal with operations like noise filtering, amplification, current-to-voltage or voltage-to-frequency conversion, and, finally, *conversion* of the analog signal to a digital signal capable of being acquired and treated by a computer.

Frequently, the reverse process is also performed: that is, a digital signal generated by the computer should be transformed into an analog signal for *control* purposes, for example, to move devices such as motors or valves, or to supply circuits for light detectors and sources, electrodes, and so on. As is shown later, A/D and D/A conversion are strongly related to each other and both are frequently implemented to perform complementary tasks in the experimental setups.

5.2. TRANSDUCERS

5.2.1. Introduction

In most cases, transducers provide low-level analog signals in response to a chemical or physical stimulus (heat, light, ionic activity, etc.). Transducers may be classified by their modes of operation as being active or passive. An active transducer generates a signal as a self-contained energy source without an external power supply. A passive transducer needs an excitation voltage or current source.

The transfer function of a transducer is the relationship between the measured property and the electrical output. Usually, it can be described by a theoretical curve and, ideally, the transfer function should be linear. In many cases, this is approximately accomplished within a certain range. Alternatively, the relationship should be linearizable via a known mathematical transformation such as a logarithm or an inversion. Thus, for instance, the resistance of a thermistor is given by the following function of the temperature:

$$R_T = A \exp(B/T), \tag{5.1}$$

where A and B are constants that must be calibrated for each thermistor unit. This function can be made linear by using natural logarithms:

$$\ln R_T = \ln A + B/T. \tag{5.2}$$

Other important features of a transducer are the response/stimulus ratio or sensitivity, noise, linearity, application range, response time, and drift.

5.2.2. Temperature Transducers

Temperature is frequently measured or controlled in many experimental situations. Some electronic devices commonly used to measure temperature are thermocouples, platinum resistors, thermistors, and several semiconductor ICs.

Thermocouples. A thermocouple is produced by welding wires of dissimilar metals together. Whenever two dissimilar metals make contact, a voltage is produced as a result of the thermoelectric effect, making this an active transducer. The temperature–voltage characteristics of a thermocouple depend on the chosen pair of metals. All thermocouples exhibit a nonlinear transfer function, but the regions of greatest curvature vary.

Thermocouples are typically used in pairs (Figure 5.1). One junction will be used as the measurement thermocouple, while the other is the cold junction. The differential voltage is proportional to the temperature difference between the two thermocouples. The thermocouple signal is small, usually below 20 mV. Differential instrumentation amplifiers can be used to reject the common mode noise picked up in the leads.

Thermopiles are multiple thermocouples connected in series, the sensitivity being proportional to the number of thermocouple junctions. The sensing junctions are in contact with the element whose temperature has to be measured.

Platinum Resistors. In a platinum resistor, the resistance of a platinum wire or band is measured as a function of temperature. Although the change of resistance with temperature is a general property of all conductors, platinum

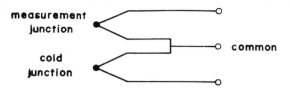

Figure 5.1. A thermocouple pair.

is the only metal used to construct resistance thermometers. In a Pt resistance thermometer, the temperature function between -180 and $630°C$ is given by the Callendar–Van Dusen equation:

$$\frac{R_T}{R_0} = 1 + \alpha\left[T + \beta\left(\frac{T}{100} - 1\right)\frac{T}{100} - \gamma\left(\frac{T}{100} - 1\right)\left(\frac{T}{100}\right)^2\right], \qquad (5.3)$$

where R_T and R_0 are the resistances at the temperature T and $0°C$, and α, β, and γ are characteristic parameters of the metal. Sometimes the following linear approximation may be used:

$$\frac{R_T}{R_0} = 1 + \alpha T, \qquad (5.4)$$

where $\alpha \approx 0.4°C^{-1}$. The typical platinum resistor, with a configuration of a coil of fine platinum wire sealed in glass, can have a value R_0 of about $100\ \Omega$.

Resistance thermometers are passive transducers that are fed with a constant current source and the voltage drop across the resistor is measured. The heat generated by the Joule effect causes some mismatch between the temperature of the resistor and the surroundings, and therefore the measuring circuit must be recalibrated when the thermal conductivity of the medium changes.

Thermistors. Thermistors are also based on the principle of the change of resistance in a predictable manner with changes in temperature, but they exhibit a much larger temperature coefficient than metals. Some thermistors have a positive temperature coefficient (PTC); that is, their resistance increases with temperature. The most common negative temperature

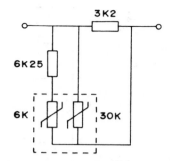

Figure 5.2. Network to improve the linearity of the resistance response for temperature measurements with thermistors.

coefficient (NTC) thermistors present a decrease of resistance with an increase in temperature.

Thermistors are made up from a sinterized semiconductor metal oxide mixture with two external connections and a plastic or glass protecting insulation. Glass-sealed thermistors built up in a rod shape are very convenient for monitoring temperature in liquids.

Most thermistors have a nonlinear response curve, but their linearity can be improved by using a circuit containing two thermistors, as shown in Figure 5.2. Since thermistors are passive transducers, a well-regulated voltage source is required. Since they have higher resistances than other resistance thermometers, the self-heating effect is not usually significant.

Semiconductor Temperature Transducers. Several semiconductor manufacturers offer devices based on the temperature dependence of the *pn* junction properties.

The LM-135, LM-335, and LM-355 are three-terminal temperature sensors, basically special zener diodes in which the breakdown voltage is directly proportional to the temperature. The two usual terminals are for power, while the third terminal is for adjustment and calibration. The transfer function is close to 10 mV/K, being accurate enough for most control applications. The LM-135 version offers uncalibrated errors of 0.5–1°C over a temperature range of −55 to 150°C. A scheme for operating an LM-355 is shown in Figure 5.3.

The current going through the LM-355 is usually set to 1 mA and the value of R has to be chosen according to the voltage source. If $V = 5$ V (TTL level supplied by digital electronic instruments), then $R = V/I = 5$ kΩ (4K7 is the nearest value of the usual standard resistor scale).

The circuit allows single-point calibration of the temperature. The calibra-

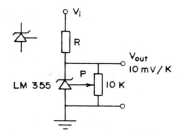

Figure 5.3. A temperature sensor circuit based on the LM-355.

tion control is obtained from the 10K potentiometer. In noncritical cases, the comparison with an ordinary mercury thermometer is sufficient. The potentiometer is adjusted until the output voltage agrees with the degrees expressed in the absolute temperature scale (e.g., 2.98 V for 298 K).

The AD-590 is another type of solid-state temperature sensor which can also operate in the −55 to 150°C range. This device is a two-terminal transducer that behaves as a current source, producing a current proportional to the temperature, typically 1 µA/K. A circuit for operating the AD-590, based on a current-to-voltage conversion by a resistor, is shown in Figure 5.4. From Ohm's law, 1 µA/K will be converted to 1 mV/K through a 1 kΩ resistor. The potentiometer is used to calibrate the system. A 1 kΩ 1% precision resistor may be more convenient for noncritical applications, since potentiometers are points of weakness in any circuit. This circuit may be used to construct a temperature alarm. For this purpose, a voltage comparator biased to the voltage that corresponds to the alarm temperature can be used.

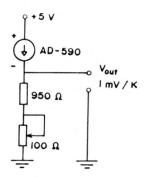

Figure 5.4. Temperature measuring circuit based on the AD-590.

5.2.3. Light Transducers

Photoelectric cells, photomultiplier tubes (PMTs), and several semiconductor chips are used to transduce UV–visible radiation. Since photoelectric cells and PMTs are extensively treated in any instrumental analysis manual, only semiconductor devices are treated here.

Photoresistors. A photoresistor is a device that changes its resistance when light is applied. In most common varieties, the resistance is very high when the photoresistor is put in the dark but drops very low under intense light.

Figure 5.5 shows a differential bridge circuit built up with photoresistors. The output voltage is

$$V_{\text{out}} = V \left| \frac{R_{\lambda 1}}{R_{\lambda 1} + R_1} - \frac{R_{\lambda 2}}{R_{\lambda 2} + R_2} \right|. \tag{5.5}$$

A differential amplifier can be connected across the output potential, V_{out}. The photoresistor may also be connected as a feedback resistor in an OA circuit, as shown in Figure 5.6. This circuit provides a low-impedance output.

Photodiodes and Phototransistors. In a photodiode, when the *pn* junction is illuminated, the level of reverse leakage current will increase due to the increase of the carrier population. The same principle applies to a phototransistor, where a collector-to-emitter current flows when the base region is illuminated. Figure 5.7 shows the basic circuit used for these sensors. The diode is normally reverse-biased and connected to a current-limiting resistance.

Both photodiodes and phototransistors are packaged in transparent cases and optimized for speed, efficiency, and low noise. They have become very useful as detectors in the visible and near-infrared regions, although

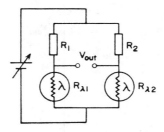

Figure 5.5. A bridge circuit with photoresistors for differential light intensity measurements.

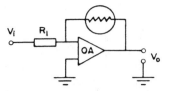

Figure 5.6. A low-impedance output circuit for light intensity measurements with a photoresistor.

some of them have been optimized to enhance their response in the ultra-violet region as well.

Photodiodes can operate as both active and passive transducers. As an active transducer, a photodiode acts as a photovoltaic cell, generating a current in the junction which may be measured directly. As a passive transducer, the photodiode response is more noisy but it is also faster and the linearity range wider.

Solid-State Imaging Detectors. The large-scale integration techniques (LSI) have been applied to the manufacturing of semiconductor light detectors. By integrating semiconductor detectors in patterns and placing a glass window rather than a plastic or ceramic block over the top, the integrated circuit can be used to detect an image—that is, intensity as a function of position.

The photosignal for each detector (picture element or pixel) is retained in the form of a charge until the output process takes place. Without the need for moving mirrors or other mechanical devices, the entire image can be acquired simultaneously. The imaging detector requires the same time for acquiring the intensity at a single element as it does for the entire image.

Because solid-state imaging devices are not capable of gain, a separate unit, called an intensifier, can be placed in front of the imaging detector to provide this gain. The best performance is obtained from a microchannel plate (µCP).

µCP operation is similar to that of a photomultiplier (PMT). A photoemissive surface faces the light, converting photons to electrons. These electrons

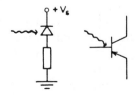

Figure 5.7. Symbols for photodiodes and phototransistors and basic circuit.

are collected by a bundle of microscopic hollow tubes that act as electron multipliers. They have a resistive inner surface so that a large voltage gradient, impressed over the length of the tube, induces multiple collisions, as does the discrete dinodes of a PMT. A gain of up to 10^4 is achieved.

In the charge-coupled device (CCD) technology, transparent electrodes are integrated over bulk silicon semiconductor material. The potential on those electrodes creates a potential source that traps the electrons produced by photoevents. As photons continue to enter the region, the charge builds in that potential source until it is full up; that is, it is saturated with about 10^6 electrons.

The source is emptied when the readout sequence takes place. By activating sets of electrodes in sequence, the charge from each detector is shifted into a newly created adjacent potential source located in an area protected from the light. Another sequence moves the charge packets stepwise until they reach a charge amplifier that sequentially converts the charge, corresponding to each detector element, into a voltage pulse at an output pin. While this readout sequence is taking place, a new photogenerated charge packet is building in the original potential source. In this way, during the period between readout sequences (the exposure time), the signal is integrated in the potential sources.

Commercial CCD arrays are produced as either linear or area arrays. In a linear array up to 4096 pixels are integrated in a linear array, usually spaced on 0.0005 inch centers. A 1000 pixel image can be read every 100 µs if a readout device that is equally as fast is available. Two-dimensional arrays have comparable spacing, but up to 128,000 pixels are incorporated in a single device.

Because of the small area covered and the front illumination employed by commercial devices, CCD arrays are not extremely sensitive, and the poor UV response (and the glass window) limits the wavelength range. However, a CCD detector array can provide a signal-to-noise ratio well over 2000:1 for multichannel measurements at high light levels. A very useful application is for detecting multiple wavelengths simultaneously in spectroscopy.

Photodiode array (PDA) detectors are built around integrated circuits with up to 1024 photodiodes linearly integrated on a single chip. Two-dimensional detectors are available as well. For each detector, a small capacitance is included and is charged to a set level. The current generated by exposure of the photodiode discharges that capacitance. When that pixel is read out, the charge on the capacitor is restored, and the output voltage is proportional to the charge required. Unlike the CCD, the readout process effectively connects a single pixel directly to the output amplifier. Consequently, the exposure period lasts from one readout to the next for that particular pixel; although all elements have equal exposure periods, the periods are shifted in time.

The UV response of the PDA elements is good, and unlike the CCD, a quartz window is used in order not to limit that response. A special version of the linear array has been produced with the spectroscopist in mind: the height of the pixels is increased to over a millimeter, vastly increasing the sensitivity.

5.2.4.   Pressure Transducers

There are two kinds of pressure transducers: those based on resistance changes, such as the strain gauge, and those that are inductive.

In a strain gauge, a wire, foil, or semiconductor element is cemented to a thin metal diaphragm. The diaphragm is flexed by the pressure, producing deformation of the element and therefore a change of electrical resistance. The linearity can be quite good, provided that the elastic limits of the diaphragm and element are not exceeded.

An example of an inductive transducer is shown in Figure 5.8. The inductive reactances of L1 and L2 are a function of the applied ac excitation frequency and the inductance of L1 and L2. When L1 = L2, the bridge is at null and V_{out} is zero. When the core attached to the diaphragm is moved, as when a pressure or force is applied, the relative inductances of L1 and L2 change, the reactances are no longer equal, and the bridge is unbalanced.

5.2.5.   Other Transducers

There are many other types of transducer of great interest to the chemist, which include selective electrodes and modified semiconductor devices that sense the activity of a number of molecules or ions. Ion selective

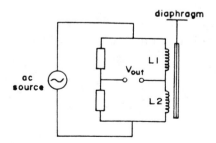

Figure 5.8. Inductive circuit for the transduction of pressure.

transducers are not treated in this book. Translation and position transducers are treated in Chapter 9.

5.3. SIGNAL CONDITIONING

5.3.1. Introduction

Frequently, the signal produced by a transducer is not adequate for its direct use and signal conditioning is required. Thus, for instance, the highest signals produced by a sensor can be on the order of 10 mV, which is too low to be fed directly to some measurement devices or A/D converters. If the full-scale range of the measurement instrument or converter is 0–5 V, precision of a measurement within the range 0–10 mV would be very poor. To improve precision, the 0–10 mV signal must be amplified up to 500-fold. The method of calculating and fabricating the corresponding amplifier circuit can be found in Chapter 2.

When the same measuring system, for example, the same computer, receives signals coming from several sensors, probably each sensor will have a different response level and will require its own amplifier circuit. This can be implemented by using a multiplexer with either a different amplifier at each input channel or a programmable gain OA at the multiplexer output. As will be shown later in this chapter, all these elements can be found on the market conveniently packed into a single data acquisition and control card.

5.3.2. The Wheatstone Bridge

The Wheatstone bridge is one of the oldest and most widely used signal conditioning circuits. It has been used for more than a century to perform resistance-to-voltage conversions. As depicted in Figure 5.9, the bridge is made up of two voltage dividers, which are called the passive (R_1 and R_2) and the active (R_{eq} and R_x) arms. The latter is used to detect or to measure resistance variations.

Since both voltage dividers are connected to the same voltage source, V_s, we have

$$V_p = V_s \frac{R_1}{R_1 + R_2} \tag{5.6}$$

and

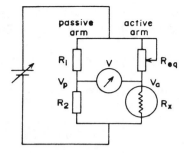

Figure 5.9. A Wheatstone bridge circuit.

$$V_a = V_s \frac{R_{eq}}{R_{eq} + R_x}. \tag{5.7}$$

When $V_p = V_a$, the voltmeter V indicates no voltage and it is said that the bridge is balanced. Under these conditions we have

$$\frac{R_1}{R_1 + R_2} = \frac{R_{eq}}{R_{eq} + R_x}. \tag{5.8}$$

A change in R_x will unbalance the bridge. It can be equilibrated by adjusting the value of the potentiometer, R_{eq}. It can be shown that the displacement of R_{eq}, which is necessary to balance the bridge, is proportional to the change produced in R_x. From Equation (5.8), we have

$$\Delta R_x = \frac{R_2}{R_1} \Delta R_{eq}. \tag{5.9}$$

Therefore, if the R_{eq} displacement scale is calibrated, the absolute value of R_x can be accurately known after balancing the bridge.

In most laboratory applications, R_x is a variable resistance transducer. Using a thermistor, a light-dependent resistor, an electrochemical conductivity cell, and so on, the Wheatstone bridge can be used as a detection system for temperature, absorbance or fluorescence, total salt concentration, and other physical or chemical properties. Moreover, the resistance R_2 can be substituted for an R_x-twin transducer, which allows differential measurements. This is a common way of avoiding unwanted effects such as a base-line drift. A drawback of the Wheatstone bridge is the dependence of the sensitivity on

R_x, which affects the linearity of the response. That dependence is reduced to a minimum by designing the bridge for a given measurement range.

5.3.3. Signal-to-Frequency Converters

If the transducer is far away from the measurement instrument and/or in a high noise environment, the signals must be appropriately conditioned and protected. Since noise and distance may affect amplitude, but not frequency in most cases, protection is accomplished by a voltage-to-frequency (V/F) conversion.

In a V/F converter, a frequency output, directly proportional to the input voltage, is produced. An adequate V/F converter, based on the use of the 458 IC from Analog Device, is shown in Figure 5.10. The frequency generated by the V/F converter can be directly input into the microcomputer for frequency counting or reconverted into a voltage for input into an A/D converter. The frequency output of a V/F converter is sometimes TTL compatible.

Between a remote sensor and the computer, the conditioned analog signal can be transmitted through a metallic wire or through an optical fiber. In highly noisy electrical environments, transmission through optical fibers is preferred. A simple TTL-compatible optical transmitter–receiver circuit is depicted in Figure 5.11. The 7400 NAND gate is employed with a driving transistor to provide an inverted pulse pattern, and the LED produces the digital light pulses. This inverted LED configuration in the transmitter is for on-line monitoring of fiber continuity. After transmission through the optical fiber waveguide, this train of inverted pulses is reinverted by the receiver.

Resistance changes can also be converted directly into a modulated frequency signal. An example is provided by the circuit of Figure 5.12. The 555 generates a TTL digital square-wave signal that acts as a carrier. The frequency of the carrier is modulated by the resistance changes in R_x.

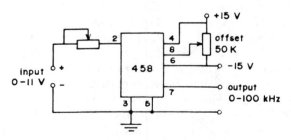

Figure 5.10. A V/F converter.

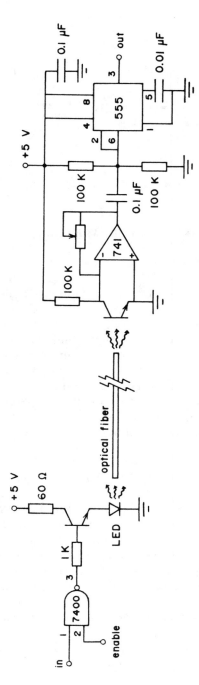

Figure 5.11. A TTL compatible optical transmitter–receiver circuit.

143

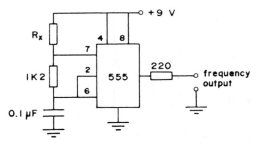

Figure 5.12. A resistance-to-frequency converter.

5.4. A/D AND D/A CONVERTERS

5.4.1. Introduction

The signals coming from transducers, either conditioned or not, are analog in nature and cannot be handled directly by microprocessors and computers. Similarly, the speed of a dc motor, the intensity of a light source, and so on must be controlled with an analog signal that the computer cannot supply. Obviously, acquisition of data by a computer requires digital encoding of analog signals, and vice versa; a computer can hardly work in the analog world, unless some form of decoding digital signals is provided. Therefore, analog-to-digital (A/D) and digital-to-analog (D/A) conversions are essential steps in any data acquisition and/or control scheme. D/A converters, which are frequently found as a part of the A/D converters, are treated first.

5.4.2. Precision in A/D and D/A Converters

Precision in A/D and D/A converters is given in bits. For a system using n digital lines or bits there are 2^n possible configurations. Since the system is capable of distinguishing each one of them, precision is proportional to $1/2^n$. For example, for a D/A converter with 8 input lines (8 bits) there are $2^8 = 256$ possible combinations of the 0 and 1 levels. For the output to be driven with the input digital number, a step of voltage is assigned to each digital unit. This means, for instance, that for an output voltage range of 0–10 V, the voltage steps have the value $10,000/256 \approx 40$ mV. The stepped output obtained in this way is shown in Table 5.1.

The converter resolution is independent of the output range. Thus, for instance, the same $(40/10,000) \times 100\% = 0.4\%$ resolution is obtained for a

Table 5.1. Output of an 8-bit D/A Converter for a 10 V Range

Binary Input	Decimal Equivalence	Output Voltage
0000 0000	0	0
0000 0001	1	40 mV
0000 0010	2	80 mV
0000 0011	3	120 mV
. . . .	. . .	. . . . . .
1111 1111	256	10 V

0–10 V range as for a 0–100 V range, or for a −24 to +24 V range. Some common values are given in Table 5.2. An 8-bit precision is more or less equivalent to the precision that is obtained with a chart recorder. Thus, with a 25 cm wide chart and a quality ruler, precision can be 1 mm, which also corresponds to 0.4%. An 8-bit precision is good enough for many applications in chemistry, since the random errors associated with the preparation of solutions are similar. Besides, 8-bit converters are inexpensive and can easily be interfaced with any μP. They can be found on the market as kits including multiplexers and other accessories.

5.4.3. D/A Converters

D/A converters are extensively used in digital signal processing, particularly for control purposes. Without decoding the digital signal, computers can only provide two levels of the positive TTL logic. Obtaining other voltages is made possible by using a D/A converter. As indicated in Figure 5.13, a digital number input to the D/A converter produces an analog output with a voltage level that is a function of the input.

Table 5.2. Precision in A/D and D/A Conversions

Number of Bits (n)	2^n	Precision (%)
8	256	0.4
10	1,024	0.1
12	4,096	0.024
14	16,384	0.006
16	65,536	0.0015

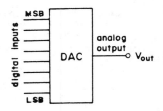

Figure 5.13. Scheme of a D/A converter.

As shown in Figure 2.45, a simple D/A converter can be created with an additive network of weighted resistors. An 8-bit D/A converter based on the 558 DAC IC is shown in Figure 5.14. An excellent variety of inexpensive D/A converters are manufactured by Ferranti. One of the most useful, the ZN425E, in combination with a few additional elements, can also work as an A/D converter.

A circuit for D/A conversion based on this 16-pin chip is shown in Figure 5.15. An internal reference source of 2.55 V is available at pin 16. This is made effective by connecting pins 15 and 16. As shown in the figure, it is convenient to use a capacitor to decouple parasitic currents. An external voltage source can also be applied to pin 15 for reference; however, because the internal source is of very high quality, no advantages are obtained. The output voltage range is approximately 0–2.5 V; however, in most applications it is necessary to implement an amplification and an output isolation or buffer steps. This is the function of the μ741 OA and associated

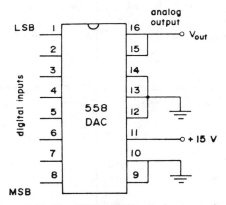

Figure 5.14. The 558 DAC IC and pin connections.

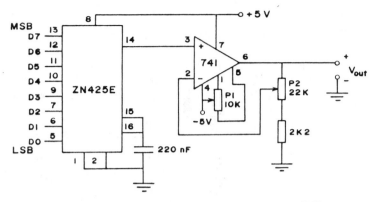

Figure 5.15. Circuit for D/A conversion using the ZN425E.

elements. The OA gain can be adjusted with the potentiometer P2, which is connected to the OA pin 6.

Ideally, the minimum D/A output is 0, which should also give 0 V at the OA output. Actually, small polarization voltages appear. The offset null potentiometer, P1, is used to compensate for these voltages. The circuit is calibrated by putting all inputs at 0 logic level and adjusting the offset null potentiometer to obtain a 0 V output. Afterward, all inputs are put at level 1, and the output potentiometer is adjusted to obtain the desired maximum value of the output. If the OA is adequately powered, the output can be as high as 25 V. The OA power supply should be at least 2 V over the maximum value of the output voltage; on the other hand, it cannot exceed 36 V, which is the maximum the OA can stand.

An important point to be considered in high-speed applications is that the output of a D/A converter does not respond instantaneously to the changes produced at the input lines. The delay is called the settling time and, for the ZN425E, is typically 1 μs.

A D/A converter circuit based on the ZN426E IC is shown in Figure 5.16. This 14-pin chip contains a low-power 8-bit D/A converter that consumes only 9 mA. The converter has an internal voltage source for reference which is available at pin 6. Connection to the reference voltage input (pin 5) requires an external load resistor and a decoupling capacitor. The converter output voltage ranges from 0 to 2.55 V. In the circuit of the figure, amplification and buffering of the output are performed by a CA 3140E OA. By feeding this OA with negative and positive supply voltages, the output voltage can swing from negative to positive values.

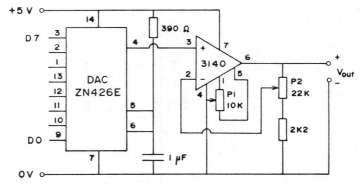

Figure 5.16. A D/A converter circuit based on the ZN426E.

5.4.4.   A/D Converters

An A/D converter transforms an analog signal, which is continuous in time, into digital words, that is, into a digital signal that is not continuous in time. Today, the huge development and wide application field of A/D converters are due to the important advantages of the digital measurement instruments which, in comparison with analog instruments, are more precise and can be programmed easier for remote control and automation purposes. On the other hand, the massive use of inexpensive and versatile µPs to substitute circuits with a wired logic also requires A/D converters to digitalize analog signals coming from all type of transducer.

As depicted in Figure 5.17, A/D converters are provided with an analog input and a codified digital output with a given number of lines or bits. Similar to D/A converters, the precision of an A/D converter depends on the number of output bits.

There are a number of techniques for A/D conversion, each one with its peculiar advantages and limitations: simultaneous or parallel encoding,

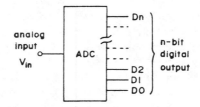

Figure 5.17. Scheme of an A/D converter.

successive approximations, sequential encoding, continuous encoding, single- and dual-slope integration, and conversion to a frequency.

Parallel Encoding. In the parallel encoding method, used by simultaneous A/D converters, the input signal voltage is fed simultaneously to one input of each of n comparators (see Figure 5.18), the other inputs of which are connected to n equally spaced reference voltages. A priority encoder generates a digital output corresponding to the highest comparator activated by the input voltage. Parallel encoding (sometimes called "flash" encoding) is the fastest method of A/D conversion, the delay time being typically less than 20 ns.

Sequential A/D Converters. A D/A converter can be used to perform sequential A/D conversions. A circuit is shown in Figure 5.19. The inputs of the D/A converter are fed with a binary counter. The counting cycle begins with all the counter outputs at low, counting goes up with the clock pulses, and finishes when all the outputs are at high, to begin again with the following clock pulse.

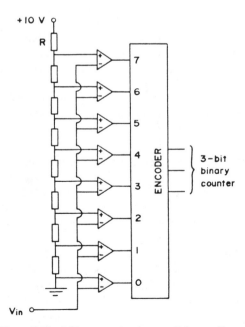

Figure 5.18. A/D conversion by parallel encoding.

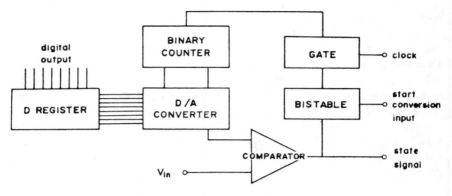

Figure 5.19. A sequential A/D converter.

The clock signal is input through a gate, which is closed when the circuit is standing by. When the start conversion input is activated by an adequate strobing pulse, the gate allows the clock pulses to arrive at the counter and counting begins. This produces a stepped ramp output at the D/A converter. This increasing output is compared with the input voltage, V_{in}. The first time the output is larger than V_{in}, the state of the output of the comparator changes. This induces the change of the bistable, the gate closes, and counting finishes. The final output of the counter is a digital number that is proportional to V_{in}. The output of the comparator can also be used to tell the computer that the conversion is completed. This is very important for two reasons: First, it is necessary to synchronize the computer with the converter. In this way, the computer will not read the converter before the end of the conversion. Second, a too high value of V_{in} prevents the count from finishing, and when the maximum possible count is reached the counter begins again from zero. Eventually, a reading operation will yield a random result. This can be avoided by checking the state of the comparator.

Slowness is the main drawback of this type of converter. Conversion time is also dependent on the value to be converted. Thus, the conversion rate is relatively high for small values but too slow for large values; for example, an 8-bit converter using this system will need 250 clock cycles to yield a value of 250.

A circuit for sequential A/D conversion based on the ZN425E IC is shown in Figure 5.20. The ZN425E contains an on-chip binary counter, and therefore it is particularly well suited for sequential A/D conversion. Pin 2 is the logic selection input of the ZN425E and, through the 1K resistor, is maintained at high. The 3140 OA is used as a comparator. The output of the D/A converter (pin 14) is connected to the OA inverting input. When the

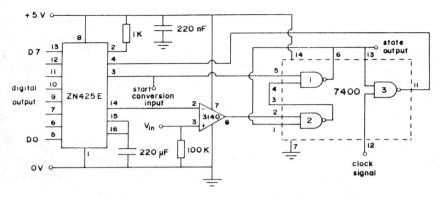

Figure 5.20. A complete circuit for sequential A/D conversion.

output of the converter exceeds the input, the OA output becomes negative and restores the bistable constituted by the NAND gates 1 and 2 of the 7400. Gate 3 is to feed the counter with the clock pulses.

The status of the output is obtained from the bistable and is at high during the conversion operation. The 8 digital outputs have no three-state capability (see Section 3.10.2) and therefore cannot be connected directly to the data bus of a computer; instead, an input port should be used.

Continuous A/D Conversion. The structure of continuous A/D converters is similar to that of sequential A/D converters, but they can reach the higher conversion rate of the simultaneous converters. In a continuous converter the counter is not set to zero after each reading. A scheme of a continuous A/D converter is given in Figure 5.21. The main difference with the sequen-

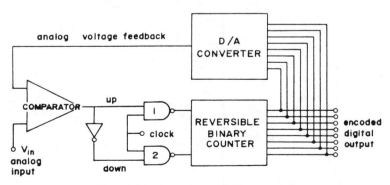

Figure 5.21. A continuous A/D converter.

tial A/D converter is the use of an up/down counter capable of performing both forward and backward counts. When a pulse arrives at the "up" input, the count increases one unit; but if the pulse comes from the "down" input, the count decreases. The upward and downward pulses are generated by the comparator. If the analog input, V_{in}, is higher than the analog feedback voltage, the count increases and vice versa. In this way, the counter output always follows the fluctuations of V_{in}.

Successive Approximations. In the popular technique of successive approximations, the output of a D/A converter is also compared with the signal to be converted. Contrary to other A/D conversion techniques, the comparison is performed after adding 1 bit to the D/A converter input. Initially, all bits are put at zero, and the MSB is added first. If the D/A converter output does not exceed the input signal, the MSB is kept at 1, but if it does, it is turned back to 0. The process is continued until the LSB has been checked. For an n-bit A/D converter, n steps are required.

As shown in Figure 5.22, this process is performed by an auxiliary register, the main register, and an adding/subtracting block that is controlled by the comparator. The auxiliary register is a shift register that is controlled by the clock. Each time a pulse arrives, all the informative contents of the auxiliary register are shifted rightward one position. In fact, only one bit of this register is at logic 1, all the other bits being always at 0. When a conversion cycle begins, the logic 1 is at the MSB position.

The main register contains the output configuration, which is fed back to the comparator through the D/A converter. The main register is loaded

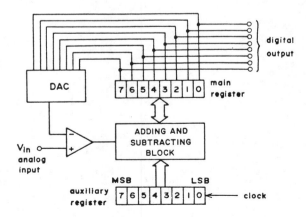

Figure 5.22. A/D conversion by successive approximations.

with the sum or subtraction of its previous contents with the binary word coming from the auxiliary register. The control of the sum/subtraction operation is performed by the signal coming from the comparator. Finally, when the displacement of the logic 1 along the auxiliary register is completed (n displacement steps), the binary word contained in the main register will represent the analog input to be measured.

The ZN427E is an A/D converter that works by successive approximations. Each conversion takes 9 clock cycles and the maximum clock frequency of this chip is 600 kHz, which means that the conversion rate can be as high as 66,000 per second. An external clock oscillator with a 1 MHz clock signal is required.

The ZN449E has an on-chip clock oscillator and requires only an external capacitor. An A/D converter circuit based on this chip is shown in Figure 5.23. Maximum clock frequency is 1 MHz. By using a 100 pF external capacitor between pin 3 and ground, the working frequency is set at 900 kHz.

Single-Slope Integration. In this technique an internal ramp generator (a current source and a capacitor) is started to begin the conversion and, at the same time, a counter is enabled to count pulses from a clock. When the ramp voltage equals the input level, a comparator stops the counter. The count is proportional to the input level.

At the end of the conversion the circuit discharges the capacitor and resets the counter, and the converter is ready for another cycle. Single-slope

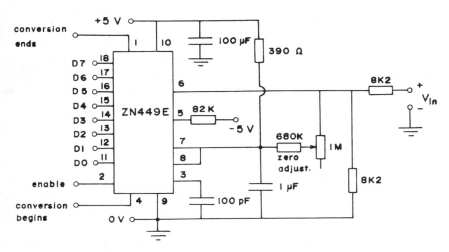

Figure 5.23. A/D converter circuit which works by successive approximations.

integration is simple, but it is not used where high accuracy is required, because it puts severe requirements on the stability and accuracy of the capacitor and comparator. The method of "dual-slope integration" eliminates that problem.

Dual-Slope Integration. This is another popular A/D conversion technique that, in comparison with other techniques, works on a rather different basis: it converts voltage in time. As plotted in Figure 5.24, both the analog voltage to be converted, V_{in}, and a reference voltage, V_{ref}, are converted in time intervals, in such a way that the ratio of the intervals equals the ratio of the voltages.

First, a current accurately proportional to the input level charges a capacitor (up to V_{cap}) for a fixed time interval. Then, the capacitor is discharged by a constant current (established by V_{ref}) until the voltage reaches zero again. The time to discharge the capacitor is proportional to the input level and is used to enable a counter driven from a clock running at a fixed frequency. The final count or digital output is proportional to the input level.

A circuit for A/D conversion based on the dual-slope integration technique is shown in Figure 5.25. The first OA works as an integrator during a fixed time, which is controlled by the control circuit. When this time has elapsed, the control circuit resets the counter to zero and changes the position of the input interruptor, from V_{in} to V_{ref}. The reference voltage, which always has the opposite sign of V_{in}, discharges the capacitor to zero.

Dual-slope integration is used extensively in precision digital multimeters, as well as in conversion modules of 10-bit to 18-bit resolution. It offers good accuracy and high stability at a low cost for applications where speed is not important.

A/D Conversion by Voltage-to-Frequency (V/F) Conversion. In this method, an analog input voltage is converted to an output pulse train,

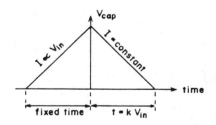

Figure 5.24. Dual-slope integration scheme.

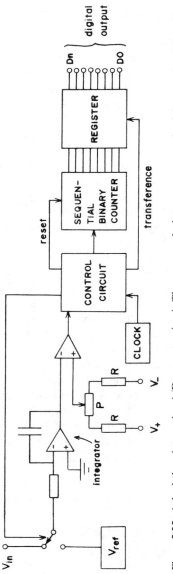

Figure 5.25. A dual-slope integration A/D converter circuit. The arrow on the interruptor means that it is opened and closed by the control circuit.

155

whose frequency is proportional to the input level. This can be done simply by charging a capacitor with a current proportional to the input level, and discharging it suddenly when the ramp reaches a given threshold.

Typically, the output frequencies of V/F converters are in the range of 10 kHz to 1 MHz. Commercial V/F converters are available with the equivalent of 12-bit resolution (0.02% precision). They are inexpensive, and they are convenient when the output is to be transmitted digitally.

If speed is not important, a digital count proportional to the average input level can be obtained by counting the output frequency for a fixed time interval. This technique is popular in moderate-accuracy (3-digit) digital panel meters.

For slowly changing analog signals (<100 Hz), the frequency-conversion technique can be modified to produce a pulse-duration converter. In this process, an analog signal (voltage, current, resistance, etc.) generates a pulse whose duration or width is proportional to the magnitude of the analog signal. If a 16-bit timer is used to measure the pulse duration, 16-bit resolution or 0.002% precision is possible.

5.4.5. Use of Analog Multiplexers with A/D and D/A Converters

Analog multiplexers (see Section 3.11.1) are frequently used to feed an A/D converter with several analog signals that are multiplexed in time. The use of a multiplexer is simpler and less expensive than the simultaneous use of several A/D converters, although the conversion rate decreases. However, there are many applications where each input needs to be read at a relatively low frequency, and to use a multiplexer is the best solution.

Besides, since analog multiplexers are bidirectional, they can be used for control purposes. Thus, for instance the output of a single D/A converter can be translated to up to 8 circuits. Since it is not possible to output 8 signals simultaneously, a sample-and-hold circuit is used at each I/O line to keep the voltage unchanged until it is renewed.

5.5. PROGRAMMING INTERFACE CARDS IN BASIC

5.5.1. Introduction

To transfer analog and digital signals among the elements of a system is an important and frequent task in automation. Microcomputers communicate with the outside world through interfaces. Many types of interface card can be found on the market. The cards are designed to be plugged into the computer standard slots, which provide connection of the card to the buses.

On the other hand, they provide 8, 16, 32, or more I/O analog and/or digital channels. The cards have a number of useful on-board circuits such as A/D converters, clocks, interrupt logics, and supporting software. The latter is usually organized as a series of useful machine language subroutines that help the user to drive the interface.

High-level languages such as BASIC or FORTRAN also provide a series of commands for the management of interface cards. Since most of these commands are seldom used outside the interfacing world, those used by interpreted and compiled BASIC dialects are explained here. However, some necessary concepts about the hardware involved are given first.

5.5.2. Device Addresses

Generally, the interface cards require a number of consecutive address locations in the I/O space of a system. The first one of these addresses is the base address of the interface. In a system, some I/O address locations will already be occupied by internal I/O and other peripheral cards; thus, to provide flexibility in avoiding conflict with other devices, the commercial cards are provided with a series of DIP switches that permit a change of their base address.

For instance, any of the hex numbers between 100 and 3FF (although some addresses may be already occupied) are available as an I/O address in the IBM PC decoded space. Such a large space allows the use of many cards in a single computer. The I/O address map of the IBM Technical Reference Manual is summarized in Table 5.3. The addresses of the table cover the standard I/O options, but if other I/O peripherals are present (e.g., hard drives, special graphic boards, and prototype cards) they will also be sharing I/O address space. Memory addressing is separate from I/O addressing;

Table 5.3. I/O Address Map of the IBM PC

Address Hex Range	Device
000–0FF	All used by internal I/O
200–20F	Game I/O adapter
278–27F	Reserved
2F8–2FF	Reserved
320–32F	Hard disk drive
378–37F	Parallel printer port
3B0–3BF	IBM monochrome display and parallel printer adapter
3F0–3F7	$5\frac{1}{4}$ inch disk drive adapter
3F8–3FF	Asynchronous communications adapter

Table 5.4. Selectable Addresses for Several Commercial Cards

Card	Selectable Addresses	Use of the Card
DASH-8[a]	Hex 100–3FF	An 8 multiplexed 12-bit A/D converter
		Four digital outputs, three digital inputs
		Three 16-bit programmable counters
DT2811[b]	Hex 200–3F8	A 16 single-ended or 8 differential multiplexed 12-bit A/D converter (unipolar–bipolar) with programmable gain
		Two 12-bit D/A converter (unipolar–bipolar)
		Eight-line digital input port
		Eight-line digital output port
FPC-011 PC ADDA-14[c]	Hex 170–1F0	A 16 multiplexed 14-bit A/D converter (unipolar–bipolar)
		Two 14-bit D/A converters (unipolar–bipolar)
FC-046 P Industrial I/O[c]	Hex 170–3C0	16 sets dry reed relay outputs
		16 sets photocouple inputs
FC-024 855 I/O card[c]	Hex 1B0–1F0	48 programmable I/O lines
		Three 16-bit counters

[a] MetraByte Corporation.
[b] Data Translation, Inc.
[c] Flytech Technology Company.

therefore, there is no possible conflict with any add-on memory. The selectable addresses for several popular cards are given in Table 5.4. As may be observed, all the addresses are within the space reserved for the I/O addresses of Table 5.3.

The commercial cards of Table 5.4 perform all their operations by means of a series of registers which form the main elements of their architecture. Each register occupies a unique address in the I/O space of the computer system. The address of a register is assigned relative to the base address, which (as explained above) is selectable within a specified range. Note that only the base address is selectable. The relative address of each register is fixed, and the address bits that select a specific register are decoded by the on-board interface logic. Table 5.5 lists all the registers of the FPC-011 PC ADDA-14 card and indicates the relative address of each register.

Table 5.5. FPC-011 PC ADDA-14 Card Registers: The Base Address is the Hex 170 (Decimal 368) or Hex 1F0 (Decimal 496)

Address	Related Function
Base + 0	Output data (0–15) to select A/D channel
Base + 1	Output Hex 00 to clear A/D register
Base + 2	Read A/D low 8-bit data (bit 0–bit 7)
Base + 3	Read A/D high 6-bit data
Base + 4	Output DA0 low 8-bit data
Base + 5	Output DA0 high 6-bit data
Base + 6	Output DA1 low 8-bit data
Base + 7	Output DA1 high 6-bit data
Base + 8	Loop back 8 times to start AD high 7-bit conversions
Base + 12	Loop back 8 times to start AD low 7-bit conversions

5.5.3. Programming Interface Cards in Interpreted BASIC

Interface applications can be programmed using different levels of complexity of the supporting software. At the lowest level, only I/O instructions are used. At intermediate levels, previously programmed subroutines, such as the machine language subroutines provided by the interface manufacturer, are called by the user's program, which avoids programming of frequently used interface operations. Finally, at the highest level, an assisted environment, such as LABTECH (Laboratory Technologies Corporation), ASYST (Software Technologies, Inc.) or HEM (Data Corporation), can be used to easily operate the interface. In BASIC, I/O instructions are the INP(X) and OUT X,Y functions. Assembly language and most other high-level languages have equivalent instructions.

INP. INP is a device I/O function that returns the byte read from an I/O port. The syntax is

INP(port)

- port is a numeric expression that has an integer value between 0 and 65,535. It identifies the hardware I/O port to read.

The INP function complements the OUT statement (see below). The INP and OUT statements give a BASIC program direct control over the hardware in a system through the I/O ports. These statements must be used carefully because they directly manipulate the system hardware.

OUT. OUT is a device I/O statement that sends a byte to a machine I/O port. Its syntax is

OUT port, data

- port is a numeric expression that has an integer value between 0 and 65,535. It identifies the destination hardware I/O port.
- data is a numeric expression that has an integer value between 0 and 255. It represents the data to be sent out of the port.

Use of the INP and OUT functions usually involves formatting data and dealing with absolute I/O addresses. Although not demanding, this can require many lines of code and requires an understanding of the devices, data format, and architecture of the card. In BASIC, the usual technique is to define the base address first, and to write or read the registers using INP or OUT, making reference to this base address. In this way, if the base address changes (the position of the switches on the board are changed), only one of the instructions of the program (the one that sets the base address) must be modified. The following example is a program written in Microsoft QuickBASIC, which reads iteratively all the A/D conversion channels of an FPC-011 PC ADDA-14 card:

```
CLS
BASEADDRESS = &H170
SERIES:
FOR CHANNEL = 0 TO 15
GOSUB INITIALIZE
LB = INP (BASEADDRESS + 2)                    'Read low 8 bit data
HB = INP (BASEADDRESS + 3)                    'Read high 6 bit data
DA = (HB − 64*(INT(HB/64)))*256 + LB)         'Final data format
PRINT "Channel = " ; CHANNEL, "Data = " ; DA
NEXT CHANNEL
GOTO SERIES
INITIALIZE
OUT BASEADDRESS + 1, 0                         'Clear A/D register
OUT BASEADDRESS + 0, CHANNEL                   'Select channel
FOR I = 1 TO 8 : A = INP (BASEADDRESS + 8) : NEXT I
                                              'Start AD high 7 bit
                                              'conversion
FOR I = 1 TO 8 : A = INP (BASEADDRESS + 12) : NEXT I
                                              'Start AD low 7 bit
                                              'conversion
RETURN
```

DEF SEG. DEF SEG is a memory statement that sets the current segment address for a subsequent PEEK function or BLOAD, BSAVE, CALL ABSOLUTE, or POKE statement. The syntax is

DEF SEG [=address]

- address is a numeric expression that corresponds to an unsigned integer between 0 and 65,535. When it is specified, it sets the beginning of the memory segment to be used by another statement or function. A value outside the range 0–65,535 produces the error message "Illegal function call." The previous segment is retained if an error occurs. If the address is omitted, the BASIC data segment is used.

DEF and SEG must be separated with a space. Otherwise, BASIC interprets the statement to mean "assign a value to the variable DEFSEG." In contrast to the BASICA dialect, in QuickBASIC, which is a compiled BASIC, the CALL and CALLS statements (see below) do not use the segment address set by DEF SEG.

DEF USR. DEF USR specifies the address where a machine language subroutine begins. That subroutine will later be called USR. The syntax is

DEF USR[n] = offset

- n can be any value from 0 to 9. It identifies the number of the USR routine of interest. If n is omitted, DEF USR0 is used.
- offset is an integer within the range 0–65,535. The value of offset is added to the value of the current segment to obtain the address where the USR routine begins.

USR. USR calls the machine language subroutine, which is indicated by the parameter arg. The syntax is

v = USR[n](arg)

- n can range from 0 to 9 and corresponds to the digit given with the sentence DEF USR to select the routine of interest. The default value is USR0.
- arg is any numerical expression or character string variable which will be the argument of the machine language subroutine. If the subroutine does not require an argument, an arbitrary argument should be given anyway. Usually, the value returned by an USR function is of the same type as the initially given (integer, string of characters, etc.). Another way of calling a machine language subroutine is CALL (see later).

As an application example let us examine how USR is used to call a machine language subroutine of the Fischertechnik general purpose interface, the INTERFAC.COM subroutine. This subroutine is very useful in simplifying the control of motors, relays, and A/D converters. The machine language program occupies the memory range from Hex FF00 to FFFF. The memory range available to BASIC is thus shortened by 256 bytes, which does not normally cause a noticeable limitation. Obviously, this part of the memory cannot be used for other purposes.

In the Interface IBM Personal Computer manual of Fischertechnik, a listing of the INTERFAC.COM assembly routine can be found. A maximum of four dc motors can be controlled by this interface. To make one of the motors turn clockwise or counterclockwise, the instruction CALL is used:

CALL M1(MCW)	'Turn motor M1 on clockwise
CALL M2(MCCW)	'Turn motor M2 on counterclockwise
CALL M3(OFF)	'Turn motor M3 off

To detect if +5 V are applied to one of the 8 digital inputs of the interface—for example, to know if a push-button interruptor is pressed—the instruction USR is used:

USR(E1)	'The function is 1 if +5 V apply at 'input E1, otherwise the output is 0

The analog inputs EX and EY of the interface are each connected to +5 V via a potentiometer. The functions USR(EX) and USR(EY) produce a value within the range between 20 and 230 according to the position of the potentiometer. Since the parameter n of USR is not specified, the default value (USR0) is used by the driver program. The program may contain other USR functions.

POKE. POKE is a memory statement that writes a byte into a memory location. The syntax is

POKE address, byte

- address is a numeric expression that returns a value between 0 and 65,535. It is an offset into the memory segment specified by the most recently executed DEF SEG statement.
- byte is a numeric expression that returns an integer value between 0 and 255. It represents the data to be written into the memory location. If the argument is a single- or double-precision floating-point value or a long integer, it is converted to a 2-byte integer.

The complementary function to POKE is PEEK. POKE must be used carefully. If used incorrectly, it can cause BASIC or the operating system to fail.

PEEK. PEEK is a memory function that returns the byte stored at a specified memory location. The syntax is

PEEK(address)

- address is a numeric expression with a value between 0 and 65,535. It is an offset into the memory segment specified by the most recently executed DEF SEG statement. If it is a single- or double-precision floating-point value or a long integer, it is converted to a 2-byte integer. The returned value is an integer in the range 0–255. The argument address is a value in the range 0–65,535.

WAIT. WAIT is a control flow statement that suspends program execution while monitoring the status of a machine input port. The syntax is

WAIT portnumber, and-expression [,xor-expression]

- portnumber is a numeric expression that has an integer value between 0 and 255. It is the number of the machine input port.
- and-expression is an integer expression that is combined with the data at the port with an AND operation. Only if the result is nonzero will the WAIT statement pass control to the next statement in the program.
- xor-expression is an integer expression combined with data from the port using an XOR operation.

The WAIT statement suspends execution until a specified bit pattern is read from a designated input port. The data read from the port is combined, using an XOR operation, with xor-expression, if it appears. The result is then combined with the and-expression using an AND operation. If the result is zero, BASIC loops back and reads the data at the port again. If the result is nonzero, execution continues with the next statement. If xor-expression is omitted, it is assumed to be zero. It is possible to enter an infinite loop with the WAIT statement if the input port fails to develop a nonzero bit pattern. In this case, the machine must be restarted manually.

The following example program line illustrates the syntax of the WAIT statement:

WAIT HandShakePort, 2

This statement will cause QuickBASIC to do an AND operation on the bit pattern received at DOS I/O port HandShakePort with the bit pattern represented by 2 (00000010).

ON TIMER. ON TIMER(n) is an event trapping statement that specifies a subroutine to branch to when n seconds have elapsed. The syntax is

ON TIMER(n) GOSUB {linelabel | linenumber}

- n is an integer expression with a value between 1 and 86,400. It is the number of seconds in the time interval.
- linelabel or linenumber is the first line of the event-handling subroutine to branch to when the time elapses.

This command provides a pseudo-interrupt. After execution of each BASIC statement, BASIC checks the timer to see if the condition $>n$ is satisfied. If it is, control passes to the subroutine; otherwise, the next line is executed. This polling of the timer is called trapping and is activated by TIMER ON. Trapping is disabled by TIMER OFF. Trapping only occurs while a BASIC program is executing (unlike a true interrupt) and can be slightly delayed by statements that require a lot of execution time.

CALL. CALL (in BASICA) is a statement that transfers control to a procedure written in another programming language. The syntax is

CALL varnum [(call-argumentlist)]

- varnum is the name of a numeric variable. The value of the variable sets the memory address where the called subroutine begins. The address must be referred as an offset with respect to the memory segment, which was defined in the last DEF SEG sentence.
- call-argumentlist are the variables or constants passed to the procedure.

As explained above, to simplify program generation, special I/O driver routines may be included together with the board, for example, DASH8.BIN for the DASH8 board of MetraByte or INTERFAC.COM for the Fischertechnik general purpose interface. These routines can be accessed from BASIC by a single line CALL statement. The routines to perform the same operations in BASIC using INP and OUT sentences would require many lines of code, would be rather slow and very tedious to program.

In order to make use of the CALL routines, they must first be loaded into memory. Care must be taken not to load them over any part of memory that is being used by another program (e.g., BASIC, print spoolers, or Disk-RAM). Otherwise, the CALL routine will not work and the PC will most likely hang up.

The simplest way to use the compiled routines provided by the card manufacturers is to select a segment of the memory map located out of the space that is used by the BASIC. For instance, for the DASH8.BIN routine of the DASH-8 card, the CALL statement may be written as follows:

```
DEF SEG = &H1700                          'Sets up load segment
BLOAD "DASH8.BIN".0                        'Loads at Hex1700:0000
OFFSET = 0
MODE% = 4
DATAIO% = X                                'Value does not matter
ERRORS% = X                                'Value does not matter
CALL OFFSET (MODE%, DATAIO%, ERRORS%)
```

where OFFSET is the address offset from the current segment of memory as defined in the last DEF SEG statement. The three variables within parentheses are the CALL parameters. On executing the CALL, the addresses of the variables (pointers) are passed in the sequence written to BASIC's stack. The CALL routine unloads these pointers from the stack and uses them to locate the variables in BASIC's data space so data can be exchanged with them.

The CALL parameters are positional. The subroutine knows nothing of the names of the variables, just their locations from the order of their pointers on the stack. Therefore, the parameters must always be in the correct order. The CALL routine expects its parameters to be integer-type variables and will write and read the variables on this assumption. If a noninteger variable is used, the routine will not work correctly.

Arithmetic functions within the parameter list parentheses cannot be performed. This is illegal and will produce a syntax error. The use of constants for any of the parameters in the CALL statement is also illegal. For example,

```
CALL DASH8 (7. 2, ERRORS%)
```

is not correct. It must be programmed as

```
MODE% = 7
DATAIO% = 2
CALL DASH8 (MODE%, DATAIO%, ERRORS%)
```

Apart from these restrictions, the integer variables can be named arbitrarily. The names used in the examples are just convenient mnemonics.

Strictly, the variables should be declared before executing the CALL. Otherwise, the simple variables will be declared by default on execution. Array variables obviously cannot be dimensioned by default and must be dimensioned before the CALL to pass data correctly, if used as a CALL parameter. The DATAIO% variable can be an array with two or more than two dimensions:

```
DIM DATAIO% (2. 100)
DEF SEG = SG
FOR I = 0 TO 100
CALL DASH8 (MODE%, DATAIO%, ERRORS%)
NEXT I
```

BLOAD. BLOAD is a file I/O statement that loads a memory-image file into memory from an input file or device. The file may contain a memory image that was stored using the BSAVE command, or it may also be a machine language subroutine that was prepared with a BLOAD at the top, compiled and converted into a file.COM using EXE2.BIN. Machine language subroutines can be loaded in an integer matrix or, also, out of the data segment of the BASIC using the BLOAD command. The syntax is

BLOAD filespec [,offset]

- filespec is a string expression containing the file specification.
- offset is the offset of the address where loading is to start.

The BLOAD statement allows a program or data, which have been saved as a memory-image file, to be loaded anywhere in memory. A memory-image file is a byte-to-byte copy of what was originally in memory.

The starting address for loading is determined by the specified offset and the most recent DEF SEG statement. If there has been no DEF SEG statement, the BASIC data segment (DS) is used as the default. If the offset is a single-precision or double-precision number, it is coerced to an integer. If the offset is a negative number in the range -1 to $-32,768$, it is treated as an unsigned 2-byte offset. If offset is omitted, the segment address and offset contained in the file (the address used in the BSAVE statement) are used. Thus, the file is loaded at the address used when saving the file.

Because BLOAD does not perform an address-range check, it is possible to load a file anywhere in memory. Care must be taken not to write over BASIC or over the operating system.

Since different screen modes use memory differently, graphic images must not be loaded in a screen mode other than the one used when they were created.

5.5.4. Programming Interface Cards in Compiled BASIC

One quick fix to improve the speed of an interpreted BASIC program is to compile it using a BASIC compiler, such as the BASICA compiler, the Turbo Basic, or the QuickBASIC. For instance, with the DASH-8 card, the reading rate of a signal for data acquisition using a FOR NEXT can be increased from about 200 samples/s to 3000 samples/s.

In compiled BASIC the instruction CALL cannot be used the same way as in interpreted BASIC. To facilitate the use of the I/O driver CALL routines in compiled BASIC or other languages (e.g., FORTRAN or PASCAL), the assembly object code files are provided. They are assembled using a Macro Assembler and are linked to other object modules from compilers. After compiling the program, the linking session is run as follows:

LINK USERPROG.OBJ + DRIVER.OBJ

The instructions that permit coding of communication processes in several compiled BASIC dialects are examined next.

CALL. CALL, in a compiled BASIC, is a statement that transfers control to a procedure written in another programming language. The syntax is

CALL name [(call-argumentlist)]

- name is the name of the procedure being called. A name is limited to 40 characters.
- call-argumentlist are the variables or constants passed to the procedure.

For example, when using the MetraByte DASH-8 card in compiled BASIC, the assembly object code file DASH8.OBJ, provided by the manufacturer, can be used. This was assembled using the IBM Macro Assembler and can be linked to other object modules from compilers:

d> LINK YOURPROGRAM.OBJ + DASH8.OBJ

When a BASIC program is compiled, the significance of the name in the CALL statement is no longer the same as in interpreted BASIC. For instance, to CALL the DASH8.BIN routine of the DASH-8 card we may use

CALL DASH8(MODE%, DATAIO%, ERRORS%)

DASH8 is not interpreted by the compiler as a variable. It becomes the public name of the subroutine to be called. Before compiling, the lines of the main program that BLOAD the DASH8.BIN routine, as well as all DEF SEG statements that control the location of the routine, must be removed. These are not required because the linker will locate the DASH8.BIN routine in memory automatically.

CALLS. In QuickBASIC it is also possible to use CALLS. The syntax is the same as in CALL. Call-argumentlist is written as:

[[{BYVAL|SEG}]argument[()]]
[,[{BYVAL|SEG}]argument[()]]. . .

BYVAL indicates that the argument is passed by value rather than by near reference (the default). SEG passes the argument as a segmented (far) address. Argument is a BASIC variable, array, or constant passed to a procedure.

CALLS is the same as using CALL with a SEG before each argument: every argument in a CALLS statement is passed as a segmented address. If the argument list of either statement includes an array argument, the array is specified by the array name and a pair of parentheses:

    DIM IntData(20) AS INTEGER
    . . .
    CALL DASH8(IntData( ) AS INTEGER)

When the CALL statement is used, the CALL keyword is optional. However, if CALL is omitted, the procedure must be declared in a DECLARE statement. When CALL is omitted, the parentheses around the argument list must also be omitted.

The SEG keyword must be used carefully because BASIC may move variables in memory before the called routine begins execution. Anything in an argument list that causes memory movement may create problems. Variables can safely be passed using SEG if the CALL statement's argument list contains only simple variables, arithmetic expressions, or arrays indexed without the use of intrinsic or user-defined functions.

CALL ABSOLUTE. CALL ABSOLUTE is a statement that transfers control to a machine-language procedure. The syntax is

    CALL ABSOLUTE([argumentlist,]integervariable)

- argumentlist are optional arguments passed to a machine-language procedure.
- integervariable is an integer variable containing a value that is the offset from the beginning of the current code segment, set by DEF SEG, to the starting location of the procedure. The integervariable argument is not passed to the procedure. The user's program may need to execute a DEF SEG statement before executing CALL ABSOLUTE to set the code segment for the called routine. Using a noninteger value for integervariable produces unpredictable results.

Arguments in argumentlist are passed to the machine-language program as offsets (near pointers) from the current data segment. Although arguments are passed as offsets, the machine-language program is invoked with a far call.

In QuickBASIC, the CALL ABSOLUTE statement is provided to maintain compatibility with earlier versions of BASIC. Mixed-language programming using the CALL statement extensions and the new DECLARE statement provide a simpler way to use assembly language with BASIC.

Also, in order to use CALL ABSOLUTE, QuickBASIC must be started with the correct Quick library, the program must be linked with QB.LIB, or the QB.QLB Quick library must be used.

5.5.5.   Other BASIC Instructions that Are Useful in Automation

BSAVE. BSAVE is an I/O statement that transfers the contents of an area of memory to an output file or device. The syntax is

BSAVE filespec, offset, length

- filespec is a string expression. It specifies the file or device on which to save the memory-image file.
- offset is a numeric expression. It is the number of bytes beyond the start of the memory segment from which to start saving.
- length is a numeric expression that has an unsigned integer value between 0 and 65,535. It is the number of memory bytes to save.

The BSAVE statement allows data or programs to be saved as memory-image files on disk. A memory-image file is a byte-for-byte copy of what is in memory, along with control information used by BLOAD to load the file.

The starting address of the area saved is determined by the offset and the most recent DEF SEG statement. BSAVE begins saving at this address. If no DEF SEG statement is executed before the BSAVE statement, the program uses the default BASIC data segment (DS). If the offset is a single- or double-precision floating-point value, it is coerced to an integer. If the offset is a negative number in the range −1 to −32,768, it is treated as an unsigned 2-byte offset.

As an example of the use of BSAVE let us examine the following:

```
DEF SEG = &HB800
BSAVE "GRAFICA.PIC", 0, &H4000
```

which is a very convenient and useful way of storing screen images during a program execution when a PC provided with a Color/Graphics monitor is used. The images can easily be retrieved with

```
DEF SEG = &HB800
BLOAD "GRAFICA.PIC", 0
```

Because different screen modes use memory differently, graphic images must not be loaded in a screen mode other than the one used when the images were created.

HEX$. HEX$ is a string function that returns a string that represents the hex value of the decimal argument expression. The syntax is

HEX$(expression)

- expression is a numeric expression that has a decimal value. The argument expression is rounded off to an integer.

VARPTR, VARSEG. VARPTR and VARSEG are memory functions that return the address of a variable. The syntax are

VARPTR(variablename)
VARSEG(variablename)

- variablename is the name of any BASIC variable, including a record variable or record element. The VARPTR function returns an unsigned integer that is the offset of the variable within its segment. The VARSEG function returns an unsigned integer that is the segment part of the variable's address. If variablename is not defined before VARPTR or VARSEG is called, the variable is created and its address is returned. When variablename is a string variable, VARPTR and VARSEG return the location of the first byte of the string descriptor.

Because many BASIC statements change the locations of variables in memory, the values returned by VARPTR and VARSEG must be used immediately. VARPTR and VARSEG are often used with BLOAD, BSAVE, CALL ABSOLUTE, CALL INTERRUPT, PEEK, and POKE or when passing arrays to procedures written in other languages. When using VARPTR or VARSEG to get the address of an array, the first element of the array must be used as the argument:

DIM A(150)
. . .
ArrAddress = VARPTR(A(1))

VARPTR cannot be used to get the address of a file's buffer. The function FILEATTR must be used to get information about a file.

5.6. DESCRIPTION OF SOME COMMERCIAL INTERFACE CARDS

As selected examples, a few popular interface cards are described next. However, it should be noted that new cards, including many from new or small companies, appear continuously on the market. New cards will usually offer better performance at a lower cost, but not always the same levels of warranty, technical assistance, and advanced software support.

5.6.1. The MetraByte DASH-8 Card

The DASH-8 card from MetraByte Corporation contains a multiplexed 8-channel A/D converter, with 12-bit precision and a maximum speed of 4000 readings/s, which is usually sufficient in many applications. It can be fitted in a half slot of an IBM or compatible PC. External connections are made through a standard 37-pin D male connector. The block diagram of the card hardware is represented in Figure 5.26. The following functions are implemented on the DASH-8:

1. An 8-channel, 12-bit successive approximation A/D converter with sample hold. The full-scale input of each channel is ± 5 V with a resolution of 2.44 mV. Inputs are single ended with a common ground. A/D conversion time is typically 25 μs and, depending on the speed of the software driver, throughputs of up to 30,000 samples/s are attainable.

2. An 8253 programmable counter timer provides periodic interrupts for the A/D converter and in addition can be used for event counting, pulse and waveform generation, and frequency and period measurement. There are three separate bit down counters in the 8253; two of them are accessible to the user. Input frequencies up to 2 MHz can be handled.

3. Seven bits of TTL digital I/O are provided, composed of one output port of 4 bits and one input port of 3 bits.

4. An external interrupt input is provided that can select any of the IBM PC interrupt levels 2-7 and allows user-programmed interrupt routines, to provide background data acquisition or interrupt driven control.

5. IBM PC bus power ($+5$, $+12$, and -12 V) is provided along with all other I/O connections on the D connector. This allows the addition of user-designed interfaces, input signal conditioning circuits, expansion multiplexers, and so on.

6. A precision $+10.00$ V reference voltage output.

Together with the card, a floppy disk that contains useful related software—such as a machine language I/O driver for control of A/D conversion, timer and digital I/O channel functions via BASIC CALL statements—is provided. The I/O driver can select multiplexer channels, set scan limits, perform software-commanded A/D conversions, interrupt driven conversions and scans, set and read the timer counter, and measure frequency and period.

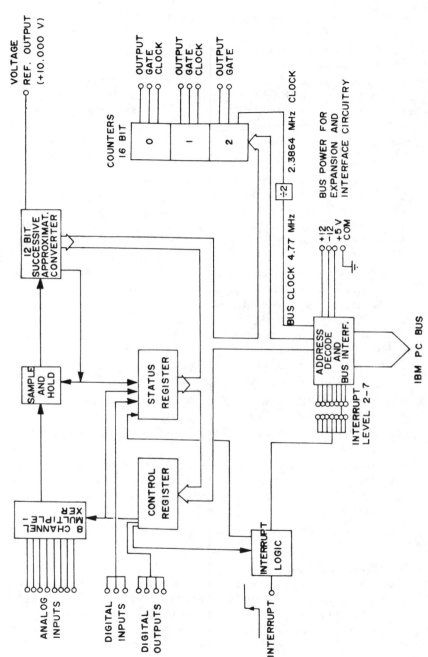

Figure 5.26. Block diagram of the DASH-8 card.

172

The capabilities of the DASH-8 may be extended by using the following expansion modules, which can be connected to the D connector:

1. Screw terminal connector board which makes the connection of all the I/O lines easier.
2. Expansion multiplexer and instrumentation amplifier EXP-16 module. This module multiplexes 16 differential inputs to a single analog output, which is suitable for connection to any of the analog input channels of the DASH-8. The EXP-16 boards are cascadable, so that up to 8 EXP-16 boards can be attached to a single DASH-8, providing a total of 128 channels. The expansion multiplexer board includes a low-drift instrumentation amplifier with preselected switchable gains of 0.5, 1, 2, 10, 50, 100, 200, and 1000. Other gains can be resistor programmed.

At the lowest level, the DASH-8 is programmed using input and output instructions (see Section 5.5.3). The use of these instructions usually involves formatting data and dealing with absolute I/O addresses. To simplify program generation, the I/O driver routine DASH-8.BIN is included in the DASH-8 software package. This can be accessed from BASIC by a single line CALL statement.

The various operating modes of the CALL routine select all the functions of the DASH-8, format and error check data, and perform frequently used sequences of instructions. An example is MODE 4, which performs a sequence of operations required to carry out an A/D conversion, check A/D status, read data, and increment the multiplexer, checking whether the upper scan channel has been reached and, if so, restoring the lower channel. A routine to perform this operation in BASIC using INPs and OUTs would require many lines of code and would be rather slow and very tedious to program.

The throughput of the DASH-8 A/D converter is a function of many interactions, but the fundamental limitation is the speed of the A/D conversion. A conversion will take a maximum of 35 µs, so that a tightly coded assembly language program, with the A/D converter operating on one channel, could easily produce a throughput in excess of 20,000 conversions/s (20 kHz). When the program is written in BASIC, the speed of the interpreter becomes the major limitation.

Interpreted BASIC takes a few milliseconds per instruction. For instance, when operating in MODE 4, about 200 conversions/s are obtained from tight loops such as

```
MOD% = 4 ; DIM X% (1000)
FOR I% = 0 to 1000
CALL DASH8 (MOD%, X%(I%), STATUS%)
NEXT
```

MODES 5 and 6 are faster, since less time is spent running through the interpreter. In fact, in interrupt MODE 6, the speed of the DASH8.BIN machine language handler and the capacity of the computer to process interrupts become the main limitations. In MODE 5, once the CALL is entered, the whole scanning cycle is performed in machine code. A maximum throughput of 4000 conversions/s can be achieved in this mode, which is the fastest obtainable operating under BASIC interpreter. The 200 samples/s rate of the FOR/NEXT loop example given above will increase to 3000 samples/s with a compiled BASIC. In a compiled BASIC program the linking session should be run as follows:

>LINK filename.OBJ + DASH8.OBJ

where DASH8.OBJ is also provided by MetraByte with the card.

5.6.2.   The Data Translation DT2811 Board Series

The DT2811 is a half-size analog and digital I/O board designed for use with the IBM/XT/AT and compatible microcomputer systems. The DT2811 performs three distinct functions: A/D conversion, D/A conversion, and digital I/O. The main difference with respect to the MetraByte DASH-8 is the possibility of software programmable amplification. The DASH-20 and DASH-16G cards of MetraByte also have this feature.

The resolution of the converters of the DT2811 takes 12 bits. Cards of this series are the DT2811-PGH, which provides a gain for A/D conversion of 1-, 2-, 4-, or 8-fold for high-level inputs, and the DT2811-PGL, which allows 1, 10, 100, and 500 gains for low-level inputs.

The A/D subsystem provides 16 single-ended or 8 differential input channels for A/D conversions. It can be configured for unipolar (0–5 V) or bipolar input ranges (+5, or ±2.5 V). The DT2811 can operate in single or continuous conversion modes. In the single conversion mode, the DT2811 performs conversion of a selected channel and stops at the completion of the conversion. The same or another channel must be selected for the next conversion operation. In the continuous conversion mode, repetitive hardware triggered conversions can be performed on a selected channel. The conversions come to a halt when the trigger is disabled. A maximum throughput of 20,000 conversions/s (20 kHz) can be achieved.

An example of use of this card is given in Table 5.6. The program, written in QuickBASIC, performs a single A/D conversion using channel 0 with a gain of 1.

Table 5.6. A QuickBASIC Program for a Single A/D Conversion Performed by the DT2811 Card

Initialization

```
CLS
BADR= &H218                    'Set BASE address
OUT BADR, &H10                 'Initialize board
FOR I= 0 TO 1000: NEXT I       'Delay
LB= INP(BADR + 2)              'Read LOW BYTE
HB= INP(BADR + 3)              'Read HIGH BYTE
GCR= &H0                       'Set GAIN = 1, Channel 0
                               '(Gain/Channel Register)
OUT BADR, &H0                  'Write MODE 0 to CSR
                               '(Control/Status Register)
```

Measure

```
OUT BADR + 1, GCR             'Load GCR (Gain/Channel
                              'Register) which starts
                              'conversion
```

Recur

```
CRS= INP(BADR)                'Read CSR (Control/Status
                              'Register)
IF CSR < 128 THEN GOTO recur  'done bit set?
LB= INP(BADR + 2)             'Read LOW BYTE
HB= INP(BADR + 3)             'Read HIGH BYTE
PRINT HEX$(HB); HEX$(LB)      'Print complete A/D data
FOR I= 1 TO 1000: NEXT I      'Delay
x$= INKEY$                    'End if any key is pressed
IF x$= "" THEN GOTO measure ELSE final  'Reload GCR
```

Final

```
CLS
PRINT "CONVERSION COMPLETE"
END
```

On the other hand, the D/A subsystem consists of two 12-bit D/A converters, each capable of operating at a throughput of up to 50 kHz. Output ranges can be independently configured to be unipolar or bipolar. An example of use of the D/A converters to generate a square wave is given in Table 5.7.

Finally, the digital I/O subsystem provides an interface for the transfer of digital data between the PC bus and one or more peripherals connected to the DT2811. It consists of two digital I/O ports, that is, port 0 dedicated to inputs and port 1 dedicated to outputs. In a data input operation, the processor reads the status of the digital I/O lines and data are transferred from an I/O device to the computer. In a data output operation, the processor sends data to the digital I/O lines, and data are transferred from the computer to an I/O device.

The DT707 optional screw terminal panel provides direct connection

Table 5.7. A QuickBASIC Program to Generate Square Waves Using the D/A Converters of the DT2811 Card

```
BADR = &H218                          'Set BASE address
LB = BADR + 2                         'DAC0 LOW BYTE address
HB = BADR + 3                         'DAC0 HIGH BYTE address
CLS

Low

OUT LB, &H0                           'Output all 0s to LOW BYTE
OUT HB, &H0                           'then to HIGH BYTE for −full
                                      'scale reading
x$ = INKEY$
IF x$ = "" THEN GOTO high ELSE GOTO final

High

OUT LB, &HFF                          'Output all 1s to LOW BYTE
OUT HB, &HF                           'then to HIGH BYTE for +full
                                      'scale reading
x$ = INKEY$
IF x$ = "" THEN GOTO low

Final

CLS
PRINT "DAC operation complete"
END
```

Table 5.8. Relative Address and Accessibility of the DT2811 Registers

Register Name[a]	Address	Accessibility[b]
A/D control/status	Base + 0	R/W
A/D gain/channel	Base + 1	R/W
A/D and DAC0 data low byte	Base + 2	R/W
A/D and DAC0 data high byte	Base + 3	R/W
DAC1 data low byte	Base + 4	W
DAC1 data high byte	Base + 5	W
DIO ports 0 and 1	Base + 6	R/W
Timer/counter	Base + 7	R/W

[a] DAC0 and DAC1 are the two D/A converters, and DIO stands for digital I/O.
[b] R = read; W = write.

points to all three subsystems of the DT2811, as well as to the external trigger and external oscillator inputs. The connection to the host computer system is made via an integral 50-conductor flat ribbon cable that plugs directly into the DT2811 board external connector.

Data Translation has developed a real-time software package, LPCLAB, which supports all the functions performed by the DT2811. LPCLAB is written in Assembly Language and provides subroutine libraries callable from a number of languages, for example, C, FORTRAN, PASCAL, and compiled and interpreted BASIC. Included routines perform single-value A/D, D/A, and digital I/O transfers, multiple-value A/D transfers to memory, full thermocouple support, on-board clock setup, and error handling.

The DT2811 performs all its operations (A/D and D/A conversions and digital I/O transfers) by means of eight registers. Each register is assigned relative to a base address, which is selectable within a specified range (see Table 5.8).

5.6.3.   The Fischertechnik Interface

This inexpensive interface is very useful for fabricating prototypes for laboratory automation. It can work with IBM compatible and other microcomputers of Amstrad, Commodore, and so on. Unlike the interfaces described above, the Fischertechnik interface is connected to the parallel printer port of the PC through a 25-pin D connector.

A scheme of the I/O lines of the interface is given in Figure 5.27. Four analog outputs for the control of motors, light sources, magnets, and so on are available. The polarity of the outputs can be reversed, for instance, to turn a motor clockwise and counterclockwise. The motors should be externally powered, with a supply of 8 V and 2 A connected to the interface. In this way, overload of the computer is avoided.

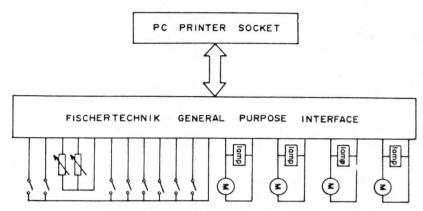

Figure 5.27. Fischertechnik interface I/O lines.

The board also contains 8 digital inputs for the detection of signals coming from push-button switches, relays, and so on. These inputs work at a TTL positive logic level and are protected against overvoltages. Finally, the board is provided with two A/D converter inputs that are designed to detect 0–5 kΩ resistance variations, such as those produced by potentiometers or many transducers.

Together with the subroutines that make driving of the interface easier, a diagnostic program to check the response of the controllable parameters is provided. The menu of the diagnostic program is given in Table 5.9.

Table 5.9. Menu of the Diagnostic Program of the Fischertechnik Interface

Type	Lines	Description of the Option or Status
Title	1	Diagnostic program
Options	2	Selected motor: keys 1–4
	3	Selected motor ccw + pulse: F1
	4	Selected motor ccw + permanent: F3
	5	Selected motor cw + pulse: F5
	6	Selected motor cw + permanent: F7
	7	Selected motor off: space bar
	8	All motors off: any other key
	9	End of program: x
Status	10	E1 E2 E3 E4 E5 E6 E7 E8 EX EY
	11	1 1 1 1 1 1 1 1 1023 1023
	12	<u>M1</u> M2 M3 M4
	13	<u>ccw</u> off off off

The option of line 2 allows selection of one of the four available motors. As status lines 12 and 13 indicate, motor 1 is selected and it is working in permanent counterclockwise mode. Once a motor has been selected, the function keys F1, F3, F5, and F7 may be used to execute motor assays either as pulses or in permanent clockwise (cw) or counterclockwise (ccw) modes. Lines 10 and 11 indicate the status of the 8 digital input lines (E1–E8) and the 2 A/D lines (EX, EY). A 1 in a digital line indicates open interruptor, and a 0 stands for closed interruptor. The values below EX and EY are proportional to the resistance between the corresponding inputs.

There are other Fischertechnik kits that are also particularly well suited to help the beginner in automation. With these kits it is easy to understand how many basic devices can be driven: for example, dc and stepper motors and optical and mechanical "end of run" and system initialization devices. Several automatic systems, such as a sampler for AAS or FIA or a multipurpose articulated arm, can easily be constructed but they will have limited performance. With the kits, the necessary subroutines and many examples written in BASIC are provided.

5.6.4. A Homemade Interface for Control of Injection Valves

We designed a prototype card that allows an IBM PC to control up to four injection valves, either for HPLC or FIA, or any other equivalent devices used in the laboratory or in the industry. One or several cards may be placed in the application slots of an IBM PC or compatible computer. The block diagram of the card is shown in Figure 5.28. It supports the hardware needed for the following:

1. Decoding the address bus, determining if the computer wants to send or to receive information to/from one of the prototype cards connected to the PC. Address bus within H300-31F (768–799).
2. Decoding the I/O READ/WRITE lines in order to determine if a datum present in the data bus has to be accepted from or given to this bus.
3. Decoding the MEMORY READ/WRITE lines in order to find out if the datum present in the address bus has to be decoded.
4. Presenting high impedance on the data bus when the card has not been selected, in order not to overload the bus. Accept or give data from/to the bus, when the card has been selected. Signals E2, E5, and E12-E14: determine which of the prototype cards has been selected. Signals E1, E3, E4, and E6–E10: data bus.

The circuit is provided with the following:

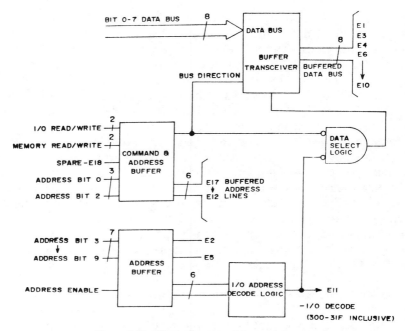

Figure 5.28. Block diagram of the prototype card.

Address Decoder. The address decoder compares the present address in the address bus (bits 0–4) with the decodified lines by means of five microswitches, which determine the address of each card. This comparison gives the CARD ENABLE signal as a result.

Control Logic of the Valves. Each bit pair of the data bus corresponds to a valve: bits 0–1 to valve 1, bits 2–3 to valve 2, bits 4–5 to valve 3, and bits 6–7 to valve 4. The even bits set the valve to a given position and the odd ones to the opposite position.

The data bus is driven toward a D flip-flop, which provides each bit to a timed monostable. This timing holds the order on the valves during a pre-fixed time to ensure the change of position. Finally, the signal coming from the monostables attack their corresponding transistors, which drive the motor relays.

Each valve has two push-button switches, which are used as end of traveling sensors. These switches inform the computer, through the data bus, about the new positions of the valves. They also allow the detection of erroneous operations: intermediate positions and failed orders.

Connection of Motors and Sensors. The scheme of connection of the motors and associated sensors (microswitches) is shown in Figure 5.29. Each motor may also be manually governed by two push-button switches. Each of these gives a POSITIVE or NEGATIVE pulse to a motor that is charged to rotate the valve in the appropriate direction. The pulse length is determined by the period of the monostables and their sign by the desired valve position.

The sensors are connected to ground and the corresponding bit is set to zero. Under normal conditions, the two bits of each valve must be different.

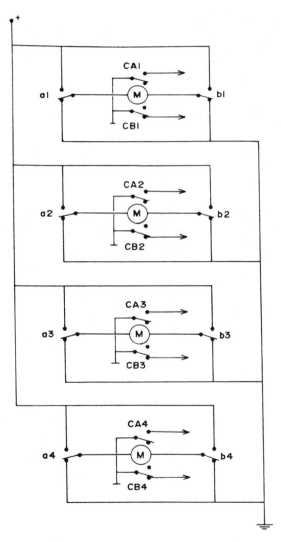

Figure 5.29. Connections of the motors (M) and sensors for control of four rotary valves: aX and bX are push-buttons; CAX and CBX are end-of-run microswitches.

Table 5.10. Subroutine to Check and Control the Valves

```
10 CLS
20 PRINT "VALVE NO. = ";
30 N$ = INKEY$ : IF N$ =" " THEN 30
40 V= VAL(N$)
50 PRINT V
60 PRINT
70 COLOR 0,15 : PRINT "O"; : COLOR 15,0 : PRINT "PEN / "; :
   COLOR 0,15 : PRINT "C"; : COLOR 15,0 : PRINT "LOSE= ";
80 V$= INKEY$ : IF V$=" " THEN 80
90 PRINT V$
100 V= V−1: IF V$= "A" OR V$= "a" THEN V= V+4
110 COD%= 2^V
120 FOR X= 1 TO 3
130 OUT 789, COD%
140 FOR Z= 0 TO 180 : NEXT Z : IF INP(789)= 255−COD% THEN 170
150 OUT 789,0
160 NEXT X
170 PRINT : PRINT
180 GOTO 20
```

If both are "1" it means that the valve has been situated in an intermediate position. If both are "0" there is a sensor failure. Check and control of the valves are carried out by a subroutine, which is listed in Table 5.10.

CHAPTER

6

COMMUNICATIONS

6.1. INTRODUCTION

Transfers of information between the elements of a system are usually performed using digital signals. For the communication to be possible, both the transmitter (sender or talker) and the receiver (or listener) must use the same language or communication system. There are several communication standards that have become popular, for example, the BCD output, the serial RS-232C interface, and the parallel IEEE-488 interface.

The communication environment determines the type of standard to be used. In serial communication, the 8 bits of each byte are converted by the transmitter to a sequence of bits; upon receipt by the receiver they are converted back to a complete byte. The alternative to serial communication is parallel communication, where the 8 bits (or more) of each byte are transferred simultaneously through a multiline cable.

In parallel communication, eight conductors to carry the byte, plus at least one conductor for electrical ground, and several more for the two devices to announce the sending and receipt of information (handshaking) must be implemented. Serial communication carries the advantage that a byte can be sent over as little as a single pair of conductors. Serial communications save cable in local systems and can also take advantage of a ready-made worldwide conductor network—the public telephone system. However, the timing and handshaking for a serial transmission can add a complication. On the other hand, parallel communication also has advantages: in the absence of parallel-to-serial-to-parallel conversions, the hardware is less complex and the transfer time is much shorter. In this chapter the RS-232C and the IEEE-488 standards are introduced.

6.2. THE RS232C SERIAL COMMUNICATION STANDARD

6.2.1. Introduction

The RS-232C (RS232C), or Recommended Standard number 232, version C, was proposed by the Electronic Industries Association (EIA), reformed to its final version, and accepted by the International Organization for

Standardization (ISO, document no. 2110). The standard was initially designed for the connection of informatic equipment (data terminal equipment or DTE, the device looks like a terminal) to a modem (modulator–demodulator, data communication equipment or DCE, the device looks like a computer) (see Figure 6.1). However, it is now widely used to interconnect equipment and peripherals without using a modem. When a modem is not used, the RS232C standard cannot be applied directly and, for this reason, the interpretation of some of the standard digital signals differs from one manufacturer of equipment to another.

6.2.2. Specifications of the RS232C Protocol

Three types of specifications may be distinguished: electrical, mechanical, and logical.

Electrical specifications are given in voltages and are:

From $+15$ to $+3$ V for a low (typical value $+12$ V)
From -15 to -3 V for a high (typical value -12 V)

In some cases, as in the Commodore VIC-20 and 64 microcomputers, the RS232C interface is only an emulation that works with TTL levels ($+5$ to 0 V). In these cases, when the output must be connected with other pieces of equipment that follow the RS232C $+/-12$ V standard, the use of a voltage adapter is required. This may be based on the MC-1488 and MC-1489 RS232C/TTL receiver/transmitter chips. A scheme of the circuit used in the authors' laboratory to connect a Crison burette with a Commodore VIC-20 microcomputer is shown in Figure 6.2.

The RS232C does not establish *mechanical specifications*; however, the use of 25-pin DB-25 connectors is customary. A connector is schematized in Figure 6.3. The connector at the computer end (DCE) should be (and usually is) a female connector. The connector at the terminal end (DTE) should be a male connector. Nine-pin connectors are also increasingly used.

The RS232C standard considers 25 *logic signals*, although in many cases, only three of them are used. The most frequently used signals are given in Table 6.1.

When the standard is applied to connect pieces of equipment directly, without a modem, the assignment of the DTE or DCE status is arbitrary. This may cause some problems when the pieces of equipment are from different manufacturers.

Figure 6.1. Original application of the RS232C standard.

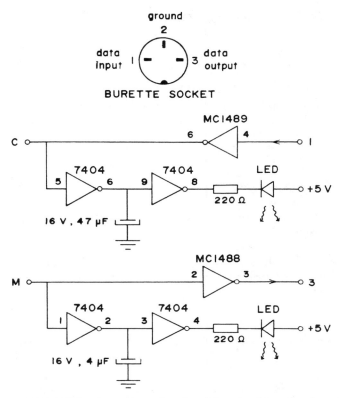

Figure 6.2. RS232C-TTL voltage adapter interface. Lines 1 and 3 work at the ± 12 V levels of the RS232C standard; C and M are connected to the VIC-20 user's port, which works at the 0–5 V levels of the TTL logic. The numbers at the IC symbols indicate pin connections.

The RS232C defines the lines with respect to the terminal equipment: that is, data are transmitted to the computer from the terminal via pin 2 and are received at the terminal from the computer via pin 3. In many applications, a simple three-wire cable, connecting only pins 2, 3, and 7, will suffice, handling the simple minimum of data transmission and ground connection.

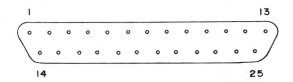

Figure 6.3. A 25-pin DB connector for RS232C interfacing.

Table 6.1. The RS232C Lines Most Frequently Used

Pin	Definition	Signal Category[a]
1	Protective ground	III
2	TxD, transmit data from DTE to DCE	I
3	RxD, receive data from DCE to DTE	I
4	RTS, request to send, data are requested to begin transmission	II
5	CTS, clear to send, the DCE is active to receive	II
6	DSR, data set ready, the DCE is active	II
7	Signal ground	I
8	DCB, carrier detect	III
20	DTR, data terminal ready, the DTE is active	II
22	Ring indicator	III
24	External clock	III

[a]Categories: I, fundamental signal required for all send–receive applications; II, control signal often used between small computers and peripherals; III, signals rarely used in small computer applications.

Some usual connection schemes are given in Figure 6.4. The choice depends on:

1. Whether handshaking is required by either or both ends.
2. Whether the ends are configured as DCE or DTE devices.

Note that where a small computer is configured to look like a computer (a DCE) but is to be used as a terminal to a second computer, both ends are configured as DCEs. Consequently, the cable must cross lines 2 and 3. Historically, a cable that connected DCE to DCE was used to test serial prototypes on a single computer by joining them and, as a result, it received the name "null modem" connection.

Thus, for instance, to connect the RS232C interface of a Perkin Elmer LS-5 fluorimeter with an IBM PC, the wiring of Figure 6.4c may be used since the fluorimeter is considered as the DTE. On the other hand, to connect many other laboratory instruments such as burettes, potentiometers, and sampling carousels, which are considered as DCEs, with a PC that is a DCE, the configuration shown in Figure 6.4b, with pins 2 and 3 crossing each other, must be used. This is also the appropriate configuration to interconnect two PCs for the direct transmission of information between them. In all these examples, pins 2, 3, and 7 suffice to implement the connection.

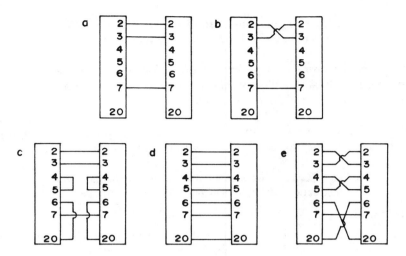

Figure 6.4. Some usual RS232C cable configurations: (a) simple three-wire, (b) three-wire null modem, (c) three-wire with loopback of handshaking, (d) full handshaking connection, and (e) null modem with handshaking.

6.2.3. Transmission Method

In serial communications, one or more devices, all of them provided with the appropriate I/O RS232C interface and connected to each other in chain fashion, communicate with the system controller. The transmission method defines how information is transferred through the serial bus. Although the RS232C standard does not define the transmission method, usually the asynchronous type with start/stop format is used. In this method, communication is initiated by a handshaking, in which the elements of the chain are requested to send back an identification number to the system controller. When instructions or data are to be sent from the controller to the elements of the chain, a start bit is sent first. The bits of a datum or instruction follow. They may be from 5 to 8, although most frequently they are 7 or 8. The next bit is for parity (odd or even) and is optional and, finally, one or two bits of stop indicate the end of the byte. To send another byte the process is repeated by beginning with the start bit. When no data or instructions are sent, line 2 (talker) goes to high and remains there.

Information is not codified for the transmission; state 0 or 1 of the line represents directly the value of each bit. This requires synchronization of the receiver with the signal, which is performed by means of the start bit.

6.2.4. Programming Serial Communications with the Commodore VIC-20 and 64 Microcomputers

The necessary commands to implement serial communications are provided with the basic package of these computers; therefore, an expansion of the available functions is not required.

OPEN. The first necessary process is to open the communication channel, establishing its working features. This is done using the following instruction.

Table 6.2. Criteria to Calculate the Controlreg Character for the Instruction OPEN in the Commodore Microcomputers

Bit Number	Weight	Function

7	128	It establishes the number of stop bits to be used. Possible values are:
		    0 = 1 stop bit
		    1 = 2 stop bits

6
5
	64
	32
		They establish the word length, that is, the number of bits that constitute each character. Possible values for bits 6 and 5, respectively, are:
		    0,0 = 8 bit word
		    0,1 = 7 bit word
		    1,0 = 6 bit word
		    1,1 = 5 bit word

4	16	Not used

3
2
1
0
	8
	4
	2
	1
		These four bits establish the communication rate. Possible values for bits 3 to 0 are:
		
		    0,0,0,0 = not implemented
		    0,0,0,1 = 50 baud
		    0,0,1,0 = 75 baud
		    0,0,1,1 = 110 baud
		    0,1,0,0 = 134.5 baud
		    0,1,0,1 = 150 baud
		    0,1,1,0 = 300 baud
		    0,1,1,1 = 600 baud
		    1,0,0,0 = 1,200 baud
		    1,0,0,1 = 1,800 baud
		    1,0,1,0 = 2,400 baud
		    1,0,1,1 = 3,600 baud (not implemented)
		    1,1,0,0 = 4,800 baud (not implemented)
		    1,1,0,1 = 7,200 baud (not implemented)
		    1,1,1,0 = 9,600 baud (not implemented)
		    1,1,1,1 = 19,200 baud (not implemented)

OPEN filenum, 2, 0, "controlreg commandreg"

- filenum is a numeric expression between 1 and 255. If its value is larger than 127, a line feed character (LF) will be sent after any carriage return (CR). The filenum permits the identification of the communication channel in all the associated instructions.
- 2 identifies the RS232C interface as the peripheral that is addressed by the OPEN command.
- 0 indicates that secondary addresses will not be used. This option is implemented as a complement in the opening of tape or disk files. The value 0 is used here only to maintain the syntax of the instruction.
- controlreg is constituted by an ASCII character which establishes the communication rate and format of the information to be transmitted or received through the communication channel. This ASCII character is a function of the value of the bits that express the character in binary. That function is the sum of the products of the weighted values of the bits and is calculated in accordance with the criteria given in Table 6.2.
- commandreg is also a numeric expression that represents an ASCII character. It is also formed by a combination of bits that indicate to the system several complementary parameters related to the communication channel configuration. The relationship between bits and parameters is given in Table 6.3.

As an example of the use of Table 6.2, let us see how the value of controlreg is calculated to implement serial communications with the potentiometers and automatic burettes of the series 517 and 738 from Crison. For a communication with 2 stop bits, a word length of 7 bits, and a rate of 2400 baud (bits per second), we have

Bit Number	Value	Weight	Product	Function
7	1	128	128	2 stop bits
6	0	64	0	7-bit word length
5	1	32	32	7-bit word length
4	0	16	0	Not used
3	1	8	8	2400 baud rate
2	0	4	0	2400 baud rate
1	1	2	2	2400 baud rate
0	0	1	0	2400 baud rate

The sum of the products gives a final value of 170. On the other hand, to fulfill the protocol specifications of the Crison 517 and 738 potentiometers and burettes, the value of commandreg is given by

**Table 6.3. Relationship Between Configuration Bits and
Complementary Parameters for the Calculation of Commandreg**

Bit Number	Weight	Function
7	128	These bits configure the type of parity control to
6	64	be performed on the received data or to be
5	32	included with the transmitted data. The possible
		values for bits 7, 6 and 5, respectively, are:
		-,-,0 = parity is disabled
		0,0,1 = odd parity
		0,1,1 = even parity
		1,0,1 = transmitted mark
		1,1,1 = transmitted space
4	16	Used to establish the directionability of the communication, which may be:
		0 = fill duplex (simultaneous)
		1 = half duplex (alternative)
3	8	Bits 3, 2, and 1 are not used.
2	4	
1	2	
0	1	It indicates the type of control signals to be used in the communication process:
		0 = control lines are not used
		1 = control lines are used

Bit Number	Value	Weight	Product	Function
7	0	128	0	Even parity
6	1	64	64	Even parity
5	1	32	32	Even parity
4	1	16	16	Half duplex
3	0	8	0	Not used
2	0	4	0	Not used
1	0	2	0	Not used
0	0	1	0	Control is not performed

which gives a total value of 112. Therefore, the serial communication channel between the Commodore VIC-20 and 64 microcomputers and the crison instruments is opened using

OPEN 2, 2, 0, CHR$(170) + CHR$(112)

Upon execution, OPEN reserves a space of the memory to contain the input and output areas. Care must be taken to codify OPEN before defining any variable or dimensioning any matrix; otherwise, the information contained in the area of the variables will be destroyed.

PRINT. To send information through the communication channel the following instruction is used:

PRINT#filenum, variable[;]

- filenum is the same number used in the instruction OPEN and specifies the peripheral that must receive the data.
- variable is an alphanumeric variable that contains the codes to be used at any moment to control the peripherals. When a semicolon [;] is added after the variable, the latter is transmitted without any additional character, whereas if it is not included, a carriage return character (ASCII 13) is automatically added after the variable.

This is an important point since some peripherals, such as the potentiometers and burettes of the examples given above, need the carriage return character to trigger the execution of the instruction sent with the variable. For instance, to fill up a burette we send

PRINT#2, "0$0IP1000"

GET. To receive information, the appropriate instruction is

GET#filenum, variable

- filenum is the same number previously used with OPEN. It specifies the peripheral that sends the information.
- variable corresponds to an alphanumeric variable where the reading process introduces a character taken from the communication channel input area. If this area is empty, the length of the variable is zero.

The program performs a loop of successive readings when the datum to be received contains more than one character. An example of how GET may be used is

GET#2, CHARACTER$

CLOSE. When all the transmission and receipt processes have finished, the communication is closed using the instruction

CLOSE filenum

- filenum is the identifier assigned to the communication channel with OPEN.

Following with the same example, we write

CLOSE 2

6.2.5. Programming Serial Communications with the Amstrad CPC-6128 Microcomputer

The necessary commands for the management of the serial interface are impressed in a ROM chip to be added to the basic configuration of this computer. The commands are presented as an extension of the standard BASIC, and all of them are coded with the symbol ":" (ASCII 124) in place of the first character. The commands are explained below.

:SETSIO. :SETSIO initializes the communication channel. The syntax is

    :SETSIO, rate tx [,rate rx [,hardware
    control [, data [, parity [, interruption]]]]]

- rate tx is an integer that indicates the rate data are sent. Possible values are 50, 75, 110, 150, 200, 300, 600, 1200, 1800, 2000, 2400, 3600, 4800, 9600, and 19200. The default value is 9600 bauds.
- rate rx is an integer that indicates the rate data are received through the interface. Possible values are the same as above.
- hardware control is an integer that indicates if a control of the status of the communication channel must be performed. The control is not performed if hardware control = 0. The default value is also zero.
- data is an integer with a minimum value of 5 and a maximum of 8. It sets the number of bits that constitute each sent or received datum. The default value is 8.
- parity is an integer that indicates the type of parity control to be performed on each sent or received datum. Possible values are:

    0, parity is not controlled in received data and a parity bit is not generated in sent data. This is the default value.
    1, a control of odd parity is performed on the received data and the data to be sent are complemented with an odd parity bit.
    2, as above, but control is performed with an even parity bit.

- interruption is an integer that indicates the number of stop bits to be sent or received with each datum. The possible values are:

0, one stop bit. This is the default value.
1, one and a half stop bits (half bit = some additional time).
2, two stop bits.

Example

:SETSIO, 2400, 2400, 0, 7, 1, 2

:**SETTIMEOUT.** :SETTIMEOUT establishes the time allowed for the data to be sent or received during the execution of an instruction of sending or receiving data. It must be codified at the beginning of the communication process. The syntax is

:SETTIMEOUT, time

- time is an integer ranging from -1 to 65,534 which indicates the time in milliseconds. If the process of sending or receiving data is not completed within the specified time, control is transferred back to the program with a timeout error message. The default value is 0. The value -1 indicates unlimited time.

Example

:SETTIMEOUT, 1000

which indicates a delay time of 1 s.

:**OUTCHAR.** :OUTCHAR sends a single character through the communication channel. The syntax is

:OUTCHAR, status, character

- status is an integer variable used to control the sending process. It must be initialized with the value 0. The possible values that the status variable may have after execution of the sentence are

0, successful process.
256, character to be sent not found.
512, unsuccessful process after timeout.

- character is the integer variable that contains the ASCII code of the character to be sent.

The variables codified in the sentences must begin with the character @ (ASCII 64). In the BASIC dialect of this computer, this is used to indicate that we want the memory address of the variable rather than its contents. In this way, the instructions will work with variable addresses instead of handling the variables directly.

Example

```
STATUS% = 0                    'Initializes the status variable
CHARACTER% = 13                'Loads the value 13 (carriage return)
                               'in the character variable
:OUTCHAR, @STATUS%, @ CHARACTER%
                               'Sends the character 13. The content
                               'of STATUS% indicates later
                               'if the process has been successful.
```

:**OUTBLOCK.** :OUTBLOCK is used to send a character block. The format is

```
:OUTBLOCK, status, block
```

- status is an integer variable that must be previously initialized. It brings back to the program information about execution of the communication. The possible final values of STATUS% are

  0, successful process.

  256, the block has not been found, or it is empty (its length is 0).

  512 + n, timeout before sending all the characters; n indicates the number of characters that have not been sent.

- block is an alphanumeric variable that contains the characters to be sent. The variables are indicated by their addresses, in the same way as shown above.

Example

```
STATUS% =  0                   'Initializes the status variable
BLOCK$ = "0S0IP100"            'Prepares the data block. This
                               'particular block is a command that
                               'fills up a burette.
:OUTBLOCK, @STATUS%, @BLOCK$
                               'Sends the block to the communication
                               'channel. Next, the contents of the
                               'status variable can be evaluated to
                               'verify the execution.
```

For the instruction that has been sent with :OUTBLOCK to be executed, a carriage return character (ASCII 13) must also be sent after the instruction. This character may be included in the alphanumeric string of the block. Let us take a look at two examples.

Example 1. The carriage return character is added to the block of data, which triggers the immediate execution of the instruction

```
STATUS% = 0
BLOCK$ = "0S0IP1000" + CHR$(13)
:OUTBLOCK, @STATUS%, @BLOCK$
```

Example 2. The data block is sent first, the communication process is evaluated, and finally, if it has been successful, the carriage return character is sent to trigger the instruction execution

```
STATUS% = 0
BLOCK$ = "0S0IP1000"
:OUTBLOCK, @STATUS%, @BLOCK$
IF STATUS% = 0 THEN ... ELSE ...
STATUS% = 0 '
CHARACTER% = 13
:OUTCHAR, @STATUS%, @CHARACTER%
```

:INCHAR. :INCHAR reads a character that has been received through the communication channel. The syntax is

    :INCHAR, status, character

where the parameters are equivalent to those of the instruction :OUTCHAR, with the difference that the variable of the parameter "character" will have already been initialized when the sentence is executed and that, after a successful process, it will contain the ASCII code of the read character. The variable assigned to "status" may contain one of the following values:

0, successful reading.

256, a variable to contain the reading character was not given.

512, timeout before successful reading.

$n*256$ ($n > 2$), reading error.

Example

```
STATUS% = 0
CHARACTER% = 0
:INCHAR, @STATUS%, @CHARACTER%
```

:**INBLOCK.** :INBLOCK acquires a data block through the communication channel. The format is

:INBLOCK, status, block

where the variable assigned to the parameter "block" must be previously initialized. It must be large enough to contain all the characters to be read. The reading process incorporates the received characters to this variable, until one of the following events occurs:

1. The specified time for the reading process is exceeded.
2. The variable assigned to "block" is full up. When this occurs no more characters can be read.
3. A signal of end of block is read. That signal can be specified by means of the instruction :SETBLOCKEND (see below).

The possible values of the variable assigned to status are

0, the variable assigned to "block" is full up.

0*256+n, n characters have been read successfully.

1*256+0, the variable that must contain the read characters has not been found, or its length is zero.

2*256+n, the character n+1 has been received out of time; only the n first characters have been read successfully.

m*256+n (m>2), an error has been produced when the nth character has been read.

Example

STATUS% = 0
BLOCK$ = " "
:INBLOCK, @STATUS%, @BLOCK$

:**SETBLOCKEND.** :SETBLOCKEND ends a reading process. The syntax is

:SETBLOCKEND, value

- value is the ASCII code of the character selected to indicate the end of the block; for example, the carriage return may be selected for this function by using

:SETBLOCKEND, 13

:CLOSESIO. :CLOSESIO closes the communication channel. The format is

:CLOSESIO, status

where the possible values of the variable assigned to status are

0, the communication channel has been closed.

512, the channel has not been closed because the data transmission area (buffer) has not been emptied within the specified limiting time; that is, a process of sending data which was initiated by a previous instruction has not yet been finished.

6.2.6.   Programming Serial Communications with the IBM PC and Compatible Personal Computers

The RS232C option is included in the standard configuration of these computers and therefore the necessary commands for the management of the interface are provided with them. A description of these commands in BASIC is given next.

OPEN COM. OPEN COM is a device I/O statement that opens and initializes a communication channel for I/O. The syntax is

OPEN "COMn:optlist1 optlist2" [FOR mode] AS [#]filenum
[LEN = registerlength]

- COMn is the name of the device to be opened. The n argument is the number of a legal communication device, such as COM1: or COM2:. When the computer is purchased with the RS232C option, the value of n is 1. When the user connects the RS232C interface to one of the slots of the computer, the position of the connections on the RS232C board should be observed to establish the value of n. The two possibilities of the IBM cards are shown in Figure 6.5
- optlist1 has the form:

  [speed] [,[parity] [,[data] [,[stop]]]].

Figure 6.5. Value of n for the IBM RS232C adapter.

The following list describes the possible options:

speed	The rate in bauds of the device to be opened. Valid speeds are 75, 110, 150, 300, 600, 1200, 1800, 2400, and 9600.
parity	The parity of the device to be opened. Valid entries for parity are N (none), E (even), O (odd), S (space), or M (mark).
data	The number of data bits per byte. Valid entries are 5, 6, 7, or 8.
stop	The number of stop bits. Valid entries are 1, 1.5, or 2.

Options from this list must be entered in the order shown; moreover, if any options from optlist2 are chosen, comma placeholders must still be used, even if none of the options from optlist1 is chosen. For example,

OPEN "COM1: ,,,,CD1500" FOR INPUT AS #1

If the data bits per byte are set to eight, no parity (N) must be specified. Because QuickBASIC uses complete bytes (8 bits) for numbers, 8 data bits must be specified when transmitting or receiving numeric data.

- optlist2: The choices for this option are described in the following list. The argument m is given in milliseconds; the default value for m is 1000.

ASC	Opens the device in ASCII mode. In ASCII mode, tabs are expanded to spaces, carriage returns are forced at the end-of-line, and CTRL+Z is treated as an end-of-file. When the channel is closed, CTRL+Z is sent over the RS232C line.
BIN	Opens the device in binary mode. This option supersedes the LF option. BIN is selected by default unless ASC is specified. In the BIN mode, tabs are not expanded to spaces, a carriage return is not forced at the end-of-line, and CTRL+Z is not treated as end-of-file. When the channel is closed, CTRL+Z will not be sent over the RS232C line.
CD[m]	Controls the timeout on the Data Carriage Detect line (DCD). If DCD is low for more than m milliseconds, a device timeout occurs.
CS[m]	Controls the timeout on the Clear To Send line (CTS). If CTS is low (there is no signal) for more than m milliseconds, a device timeout occurs.
DS[m]	Controls the timeout on the Data Set Ready line (DSR).

<table>
<tr><td></td><td>If DSR is low for more than m milliseconds, a device timeout occurs.</td></tr>
</table>

LF	Allows communication files to be printed on a serial line printer. When LF is specified, a line feed character is automatically sent after each carriage return character. This includes the carriage return sent as a result of the width setting. Note that INPUT and LINE INPUT, when used to read from a COM file that was opened with the LF option, stop when they see a carriage return, ignoring the line feed.
OP[m]	Controls how long the statement waits for the open to be successful. The parameter m is a value in the range 0 to 65,535 representing the number of milliseconds to wait for the communication lines to become active. If OP is specified without a value, the statement waits for 10 s. If OP is omitted, OPEN COM waits for 10 times the maximum value of the CD or DS timeout values.
RB[n]	Sets the size of the received buffer to n bytes. If n is omitted, or the option is omitted, the current value is used. The current value can be set by the /C option on the Quick-BASIC or BC command line. The default is 512 bytes. The maximum size is 32,767 bytes.
RS	Suppresses detection of Request To Send (RTS).
TB[n]	Sets the size of the transmit buffer to n bytes. If n is omitted or the option is omitted, the current value is used. The default size is 512 bytes.

The options from the list above can be entered in any order, but they must be separated from one another by commas. For CS[m], DS[m], and CD[m], if there is no signal within m milliseconds, a timeout occurs. The value of m may range from 0 to 65,535, with 1000 as the default value. The CD default is 0. If m is equal to 0 for any of these options, the option is ignored. If the CS option is specified, the CTS line is checked whenever there are data in the transmit buffer. The DSR and DCD lines are continuously checked for timeouts if the corresponding options (DS, CD) are specified.

- mode is one of the following string expressions:

OUTPUT	Specifies sequential output mode
INPUT	Specifies sequential input mode
RANDOM	Specifies random access mode

If the mode expression is omitted, it is assumed to be random access input/output. The filenum is the number used to open the file. The OPEN COM statement must be executed before a device can be used for communication using an RS232C interface.

- If the device is opened in RANDOM mode, the LEN option specifies the length of an associated random access buffer. The default value for length is 128. Any of the random access I/O statements, such as GET and PUT, may be used to treat the device as if it were a random access file.

The OPEN COM statement performs the following steps in opening a communication device:

1. The communication buffers are allocated and interrupts are enabled.
2. The Data Terminal Ready (DTR) line is set high.
3. If either of the OP or DS options is nonzero, the statement waits up to the indicated time for the Data Set Ready (DSR) line to be high. If a timeout occurs, the process goes to step 6.
4. The Request To Send (RTS) line is set high if the RS option is not specified.
5. If either of the OP or CD options is nonzero, OPEN COM waits up to the indicated time for the Data Carrier Detect (DCD) line to be high. If a timeout occurs, the process goes to step 6. Otherwise, OPEN COM has succeeded.
6. The open has failed due to a timeout. The process deallocates the buffers, disables interrupts, and clears all the control lines.

In comparison to the CS, DS, or CD options, a relatively large value for the OP option should be used. If two programs are attempting to establish a communication link, they both need to attempt an OPEN during at least half of the time they are executing. Any syntax errors in the OPEN COM statement produce an error message that reads "Bad file name."

Example. For the Crison 517 and 738 instruments (see above) the following values of the parameters should be used: rate, 2400; parity, "E"; data, 7; and stop bits, 2. Since only three lines (2, 3, and 7) are implemented in the burettes of the 738 series, control signals cannot be used, and the command must be written as

OPEN "COM1: 2400, E, 7, 2, RS, CS, DS, CD" AS #1

PRINT. The instruction PRINT is used to send information through the channel. Syntax is

PRINT # filenum, characters [;]

- filenum is a numeric expression with the same value adopted for n in the sentence OPEN "COMn . . . .
- characters may be an alphanumeric constant or alphanumeric variable containing the characters to be transmitted.
- [;] is optional. If not included, upon execution of the PRINT sentence, the computer sends a carriage return character (ASCII 13) after the specified characters. If included, the carriage return character is not sent. This may allow the user to cancel an instruction that has been sent before its execution.

Example 1. To send an order for the identification of the devices in a serial chain ("N"), together with a carriage return to trigger immediate execution of the order, we write

PRINT #1, "N"

Example 2. To send the order of filling up the burette, the following can be used:

DATA$ = "0S0IP1000"	'The string containing the order 'is assigned to a variable.
PRINT #1, DATA$;	'The order is sent, but it is not 'executed due to the 'semicolon.
CARRIAGETURN$ = CHR$(13)	'Assigns the carriage return 'character to a variable.
PRINT #1, CARRIAGETURN$;	'Sends a carriage return. The 'semicolon prevents a second 'carriage return from being sent.

The information received by the computer through the communication channel is stored in a reserved memory area known as buffer. The size or capacity of this area may be changed using the correct instructions, the default value being 255 characters. Also, with the appropriate software, the information contained in the buffer can be retrieved. To make this process easier, the status of the buffer must first be established. This is done by using the BASIC sentences LOC, LOF, and EOF, which are given below.

LOC. LOC is a file I/O function that returns the current position within the file. The syntax is

LOC(filenumber)

- filenumber is the number used in the OPEN statement to open the file.

For a communication device, LOC(filenumber) returns the number of characters in the input queue waiting to be read. If that number is higher than the buffer capacity, the latter is the number returned (255 if the default value is used).

With random access files, the LOC function returns the number of the last record read from or written in the file. With sequential files, LOC returns the current byte position in the file, divided by 128. With binary mode files, LOC returns the position of the last byte read or written.

The value returned depends on whether the device was opened in ASCII or binary mode. In ASCII mode, the low-level routines stop queuing characters as soon as end-of-file is received. The end-of-file produces an error message that reads "Input past end of file." In binary mode, the end-of-file character is ignored and the entire file can be read.

LOF. LOF is a file I/O function that returns the length of the named file in bytes. The syntax is

LOF(filenumber)

- filenumber is the number used in the OPEN statement.

When used on a device opened as a file with the statement OPEN COM, the LOF function returns the number of free bytes in the output buffer. It is equivalent to the difference between the value obtained with LOC and the total size of the area. When a file is opened in any mode, the LOF function returns the size of the file in bytes.

EOF. EOF is a file I/O function that tests for the end-of-file condition. The syntax is

EOF(filenumber)

- filenumber is the number used in the OPEN statement.

EOF informs the user if there is still one (or more) character in the communication buffer. In this case, EOF returns 0, but if the buffer is empty it gives −1.

INPUT$. To retrieve the information contained in the buffer, the following sentence is be used:

variable = INPUT$ (length, [#]filenumber)

- variable is the name of an alphanumeric variable where the characters to be retrieved from the buffer are to be loaded.
- length is the number of characters to be read and loaded in the variable.
- filenumber is the identifier assigned to the communication channel with OPEN.

Example 1

IF NOT EOF(1) THEN DATA$= INPUT$(LOC(1),#1)

If the buffer is not empty, the indicator associated with EOF is activated and the sentence after THEN executed. The process of acquisition of data will be controlled by the function LOC and thus, upon execution, all the characters contained in the buffer will be loaded in the variable DATA$.

Example 2

```
WHILE EOF(1)
WEND
```

A waiting loop controlled by the status of the buffer is performed. The loop is repeated until the buffer is not empty. An example similar to this one is

```
WHILE LOC(1) < MINIMUM%
WEND
```

The loop is now repeated until the buffer contains a number of characters equal to or larger than the value of the variable MINIMUM%. In both cases, care must be taken to avoid the generation of a closed loop, which will block the system. For this purpose, it is convenient to introduce keyboard or time controls, as shown below:

```
KEYBOARD$= " "
    WHILE EOF(1) AND KEYBOARD$= " "
WEND
IF EOF(1)
    THEN ...                    'Keyboard escape
    ELSE ...                    'Data are present in the buffer
```

When any key of the keyboard is pressed, its corresponding value is loaded in the variable KEYBOARD$, and thus the program control escapes from the loop. Another possibility is

```
TIMEINT= TIMER + DELAY
   WHILE EOF(1) AND TIMER < TIMEINT
WEND
IF EOF(1)
   THEN...                              'Timeout escape
   ELSE...                              'Data are present in the buffer
```

In this case, TIMER, a function of the internal clock of the system (see below), is used. With the variable DELAY, a given number of seconds are added to TIMER. The loop ends when any character is loaded in the buffer, or when the seconds indicated in DELAY have elapsed.

Example 3

```
INDEX$= 0
WHILE NOT EOF(1)
   INDEX%= INDEX% + 1
   CHARACTER$ (INDEX%)= INPUT$ (1, #1)
WEND
```

In this example, the alphanumeric string variable CHARACTER$(X) is used to retrieve the information contained in the buffer. Each element of the string stores one character and, when the buffer is empty, the variable INDEX% is used to establish the number of characters that have been read.

The BASIC functions and statements that are frequently used in relation to the control of delay loops are given next.

ON COM. ON COM(n) is an event-trapping statement that specifies a subroutine to branch to when characters are received at communication port n. The syntax is

ON COM(n) GOSUB {linelabel | linenumber}

- n is an integer expression that indicates one of the serial ports, either 1 or 2.
- linelabel or linenumber is the first line of the event-handling subroutine.

If the program receives data using an asynchronous communication adapter, the BASIC command-line option /C can be used to set the size of the data buffer.

TIMER. TIMER, as a function, returns the number of seconds elapsed since midnight. The syntax is

TIMER

TIMER, as a statement, is an event-trapping statement that enables, disables, or inhibits timer event trapping. The syntax is

TIMER ON
TIMER OFF
TIMER STOP

- TIMER ON enables timer event trapping by an ON TIMER statement. While trapping is enabled, a check is made after every statement to see if the specified time has elapsed. If it has, the ON TIMER event-handling routine is executed.
- TIMER OFF disables timer event trapping. If an event takes place, it is not remembered if a subsequent ON TIMER is executed.
- TIMER STOP disables the timer event trapping. However, if an event occurs, it is remembered and the ON TIMER event-handling routine is executed as soon as trapping is reenabled with TIMER ON.

TIME$. TIME$ is a function that returns the current time from the operating system. The syntax is

TIME$

An eight-character string is returned, in the form hh:mm:ss, where hh is the hour (00–23), mm is the minute (00–59), and ss is the second (00–59).

CLOSE. When the data transmit/receive process has ended, the communication channel is closed with

CLOSE [[#]filenumber]

If filenumber is not given, all the open files (including the open disk files and communication channels) will be closed.

6.3. THE IEEE488 INTERFACE

6.3.1. Introduction

The IEEE-488 (IEEE488 or "IE cube") interface was developed around 1970 by Hewlett-Packard and in 1975 became adopted by the Institute of Electrical and Electronic Engineers (IEEE) as digital interface for programmable instrumentation. It is referred to by several names, for example, HP-IB (Hewlett-Packard interface bus) and GPIB (general purpose interface bus). Since the IEEE488 can interconnect up to 15 external devices, it is regarded as the IEEE488 bus.

In comparison to the RS232C serial interface, the IEEE488 is a more rigidly specified standard. It defines the electrical specifications as well as the cables, connectors, control protocol, and messages required to allow information transfer between devices.

6.3.2. Controllers, Listeners, and Talkers

Devices on the bus are categorized according to three types: controllers, listeners, and talkers.

Several controllers may be connected to the same bus, one of them having the special attribute of being the system controller. At any one time there is only one active controller on the bus, which can be the system controller or another one. The system controller allows a specific talker to communicate with one or several listeners. Only one talker is active at any one time, but several listeners may be active simultaneously. The system controller determines which device talks and when it can talk. The system controller can also appoint a replacement, which then becomes the active controller. It can define which devices will listen, if the information is not required by all the devices on the bus.

Listeners are primarily data receivers. Printers, plotters, and programmable power supplies are all examples of listeners. Listeners may occasionally talk when they are asked for status information. Devices that are typically listeners may have a "listen always" or "listen only" switch, next to the device address switches.

Talkers are primarily data transmitters. Digital voltmeters, analog-to-digital converters, and digitizers are all examples of talkers. Talkers must occasionally listen, so that they will know when to begin and end transmission and when to change their status. Unlike listeners, there can only be one talker at a time. When a device is addressed to talk, it typically remains configured to talk until it receives an interface clear (IFC), the talk address of another instrument, its own listen address, or an universal untalk command. Talkers may have a "talk always" or "talk only" switch. This switch allows the talker to communicate with a listen only device without the need for a controller.

Upon power-up, the system controller is active. Most often, this system controller is the main computer. All remaining controllers become active only when control is passed on to them. The active controller wakes up inactive listeners by asserting attention and sending bus commands that specify the new talker and listeners. It is the assertion of attention that establishes the fact that the data on the bus represent an interface command. When attention is not asserted, the data on the bus represent a message from a talker to a listener.

While most of the data transfer is routine between talkers and listeners, occasionally a special command is required to perform a specific function. Thus, the IEEE488 makes a hardware handshake during every data transfer to ensure that no device on the bus misses any information. This protocol allows the IEEE488 to accommodate both fast and slow listeners receiving data from the same source. Typically, instrumental data rates are in the 1–10 kbytes/s range. In actual practice, however, the rate of data transfer on the IEEE488 bus is governed by the slowest active listener.

6.3.3. Device Addresses

The controller has the task of locating and communicating with all the devices on the bus. To do this, each device on the bus has an address by which it is identified. Once addresses are assigned, the controller can locate and define listeners, talkers, or other controllers. Each address must be unique and is usually assigned by external or internal switches or by internal jumpers. Figure 6.6 shows a typical address switch set for device address number 6 ($2^1 + 2^2 = 6$), listen always, and service request disabled.

The IEEE488 standard does not allow a device address of 31 (30 is the highest). However, the address 31 is used on some devices for diagnostic purposes.

If a device address is changed while the power is on, it is often necessary to turn the power off and then on again, for the address change to be recognized by the system controller.

Secondary addresses are normally used to access additional operating modes on a single device. These extended operating modes are not defined by the IEEE488 protocol, so they must be determined from the operating manual of the device. Any single device may have up to six secondary addresses. Secondary addresses, sent as ASCII characters, do not distinguish between talkers and listeners. A device will recognize a secondary address if it is currently addressed to talk or listen.

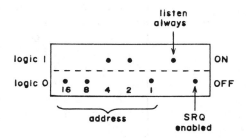

Figure 6.6. A typical address switch.

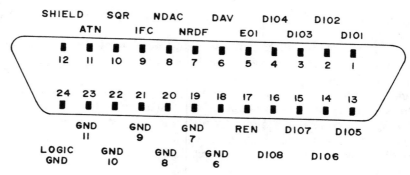

Figure 6.7. An IEEE488 24-pin connector.

6.3.4. The IEEE488 Bus

The interface IEEE488 is constituted by 16 signal lines: 8 for data, 5 for control, and 3 for transfer. The IEEE488 uses negative logics; that is, the lines become active when driven to low levels.

The 8 *data lines* (DI01–DI08, see Figure 6.7) are used to transfer data to and from all instruments connected to the bus. Bits are transferred in a parallel fashion as bytes, whereas bytes are transmitted in a serial fashion. A byte can represent any type of I/O, control, address, or status information.

The 5 *control lines* are responsible for an orderly flow of information. The function of each line is:

ATN (attention)	It is asserted by the active controller. When active, all devices are required to monitor the data lines, which carry control/address information. When inactive, all information on the data lines is interpreted as data.
EOI (end or identify)	It is activated by a talker to identify a datum as the last one in a series of transmitted data bytes. It can also be used with the ATN line to effect a polling sequence.
IFC (interface clear)	It is typically used in error/fault situations to terminate all bus activity. Only the system controller can assert the IFC command. When active, all devices assume an inactive status, talkers and listeners become unaddressed, and control is returned to the system controller, which becomes the active controller.
REN (remote enable)	Allows instruments on the bus to be programmed by the active controller. It is

activated only by the system controller and monitored by instruments capable of being remotely controlled. Any device that is addressed to listen while REN is true may be programmed by the active controller.

SRQ (service request)   This line is made active by a particular device requesting the attention of the controller and an interruption of the ongoing sequence of events.

The 3 *transfer lines* control the flow of information over the data lines. They operate in a three-wire interlocked handshake mode, which allows the asynchronous transfer of data bytes between devices on the bus. As previously stated, the speed is dictated by the talker or by the slowest active listener. Transfer functions are:

DAV (data valid)   It is activated by a talker when valid data are present on data lines. When inactive, information on the data lines is regarded as invalid data.

NRFD (not ready for data)   It is activated by a minimum of one listener on the bus to indicate its inability to receive data. When inactive, all potential listeners are ready to receive data.

NDAC (not data accepted)   It is activated by a minimum of one listener to indicate that it has not yet accepted the data. When this line is inactive, it indicates that all listeners have accepted the data.

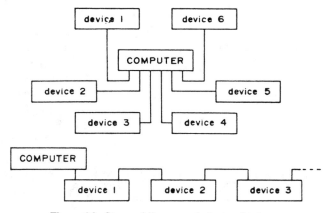

Figure 6.8. Star and linear or chain topologies.

The devices are connected to the IEEE488 bus by means of a 24-pin connector. Therefore, there are eight remaining lines which are not used, and one of these may be designated as the cable shield ground. A scheme of the connector showing the pin functions is given in Figure 6.7.

Connections and cabling can be made following a star or a linear configuration (Figure 6.8). The star cabling topology minimizes worst-case transmission path lengths but concentrates the system capacitance at a single device. The linear cabling topology produces longer path lengths but distributes the capacity load. Combinations of star and linear cabling configurations are also acceptable.

6.3.5. Interface Functions

There are 10 interface functions specified by the IEEE488 standard. These functions establish how the device will operate in stand-alone or system environments. It is not necessary for a device to implement all these functions; only a subset of them may be implemented for reasons of economy. Therefore, most devices will respond to only a limited number of commands. Some devices will list the functions that they implement next to the IEEE488 connector. The 10 interface functions are:

SH (source handshake)	It provides a device with the ability to transfer messages from a talker to one or more listeners.
AH (acceptor handshake)	It provides a device with the ability to properly receive data from a talker.
Tx, TEx (talker or extended talker)	It allows a device to send data and status messages when it is addressed to talk by the controller. A single-byte talk address establishes a talker; a two-byte address establishes an extended talker.
L, LE (listener or extended listener)	It allows a device to receive data and status messages when it is addressed to listen by the controller.
SR (service request)	It allows a device to asynchronously request service from the controller. The service request will remain active until the device is serviced. If the service request consists of a multibyte transfer, service request will continue to be asserted until all bytes are read.
RL (remote/local)	It allows the device to select between two modes of operation: remote

	(via IEEE488) or local (front panel controls).
PP (parallel poll)	Parallel polling is a method of simultaneous checking on the status of up to eight instruments on the bus. When a parallel poll is initiated, each device returns a status bit via an assigned data line. The PP function allows a device to use one bit to identify its requirements for service in response to a parallel poll command from the controller.
DC or DCL (device clear)	It allows a device to be cleared or initialized to a status defined by the device.
DT (device trigger)	It allows a device to be triggered or synchronized with other devices.
C (controller)	It allows a device to send bus commands and data, and to address devices to talk and listen. It may also initiate serial or parallel polling. Any device on the bus can be a controller, but there may be only one active controller at a time. Only one controller can be the system controller.

6.3.6. Programming of IEEE488 Commercial Cards

In order to facilitate programming to the users, the IEEE488 commercial cards are usually provided with some assembly subroutines that are resident in ROM or EPROM chips. This resident firmware converts some user's commands and data strings to specific control codes for the IEEE488 controller chip. It also passes back received data and interface status conditions to the user's program. Received information can be used directly in the user's program, and the received status codes are useful in determining various interface operating conditions or detecting syntax errors in the commands.

From the user's point of view, assembly language subroutines can be considered as high-level language extensions (e.g., of BASIC). These extensions consist of statements like INITIALIZE, TRANSMIT, RECEIVE, SEND, DEVICE, and ENTER, and they allow the PC to execute statements that are similar to the BASIC ones, but they are executed much faster because they are written in assembly language. The statement names only represent address offsets, and these offsets can be given any preferred name (usually a mnemonic one).

The PC<=>488 interface from Capital Equipment Corporation (CEC) is used below as an example to describe some common IEEE488 subroutines and to show how to make practical use of them. The PC<=>488 card provides a series of versatile and easy to use machine language subroutines, packaged in a ROM. This card is compatible with the IBM PC/XT and AT series and clone computers and has been used in the authors' laboratory with excellent results to transfer programs from HP-85/86 to IBM PCs. After correcting the programs with a word processor (e.g., Word Perfect), they have successfully been run with a Microsoft BASIC. Also, we have used the card to interface Keithley multitesters operating in the data logger mode. This is a convenient way of implementing a low-cost data logger. A limitation of these systems is the relatively low rate for acquisition of data.

Very little code is required to program the PC<=>488. The code required typically represents a small percentage of the lines in a program. Together with the PC<=>488 card, a floppy disk containing several utilities is provided. These utilities are very useful in acquiring practice on the IEEE488 interface standard and include software for transfer of files between a printer or a plotter and a computer, and between two computers. The PC that receives the file looks like a printer to the computer sending it.

An assembly language subroutine is called from a BASIC program with the USR function or the CALL statement. The latter is usually recommended because it is more readable and can pass multiple arguments (as CALLS). Besides, the CALL statement should be used to access subroutines that were written using Microsoft FORTRAN calling conventions. The syntax of the CALL statement is

CALL variable name (argumentlist)

where variable contains the offset into the current segment, which is the starting point in memory of the subroutine being called. The current segment must have been previously defined by a DEF SEG statement. Therefore, calling an assembly language routine requires three steps:

1. The location of the routine must be defined using a DEF SEG statement. This defines the current segment location of the routine.
2. The routine must be located within the segment as defined by an offset variable.
3. The routine is executed by using a CALL statement.

In the PC<=>488, the assembly language routines are located in memory at the segment decoded by the address switch on the interface. For example, if the switch decodes segment "C", then the correct statement to use is DEF SEG= &HC000.

Every called routine must also define its start address with an offset from the current segment. The offsets determine where the program will branch, and an improper location can cause the microprocessor to ignore all inputs. This is true for all BASIC CALL and USR subroutines.

The DEF SEG statement and all offset addresses are assigned with a single BASIC statement and never require reassignment within the program.

Each routine also specifies parameters that it must receive from the BASIC program, or parameters that it passes back to the BASIC program. These parameters are shown in parentheses following the program offset for each statement.

There are some limitations in BASIC that have placed restrictions on the called routines:

1. The order, number, and type of variables passed to the routines must be exactly as shown for each routine. This is because BASIC only passes pointers to variables and does not provide a variable-type identifier or delimiter to indicate when a pointer ends and another program statement begins.
2. The BASIC interpreter and compiler have different string variable requirements. Therefore, the entry points for routines with string variables (TRANSMIT, RECEIVE, SEND, and ENTER) must change when moving code from the interpreter to the compiler.
3. Passed values must be variables and cannot be constants.
4. BASIC may change a source code string if a subroutine changes the contents of the passed string. This situation can be avoided by assigning a value to the string argument before calling an assembly language routine.

6.3.7. Assembly Language Subroutines of the CEC PC<=>488 Card

Some of the most interesting subroutines of the PC<=>488 card, together with the necessary statements and commands to manage them, are described next.

INITIALIZE. INITIALIZE initializes the interface and, optionally, the bus. The syntax to call it is

CALL offset (addr%, level%)

- offset represents the offset into the current segment defined by the last DEF SEG instruction. For the INITIALIZE routine it must be set to 0 and assigned any name (e.g., INIT).
- addr% is the address of the interface card and must be in the range of 0–30.
- level% is the initialization level, which can have the following values:

level% = 0, the interface becomes the system controller.

level% = 1, the interface becomes the system active controller.

level% = 2, the interface becomes a controller.

Example

```
DEF SEG= &HC000              'Assigns segment address
INIT= 0                      'Assigns offset address
ADDR%= 21                    'Assigns controller address
LEVEL%= 0                    'Assigns system controller
CALL INIT(ADDR%, LEVEL%)     'Initializes the interface
```

TRANSMIT. TRANSMIT transmits any sequence of commands and/or data on the bus. The syntax is

CALL offset (comdata$, status%)

- offset represents the offset into the current segment defined by the DEF SEG instruction. It is 3 for the BASIC interpreter and 30 for the BASIC compiler.
- comdata$ is any sequence of commands and/or data. Commands within the string are separated by one or more spaces.
- status% is returned after the call with one or more bits set under the following conditions:

0, illegal syntax.

1, tried to send when not a talker.

2, a quoted string or END command found in a LISTEN or TALK list.

3, bus timeout or no acceptors.

4, bad command mnemonics.

Example

```
DEF SEG= &HC000              'Assigns current segment
TRANSMIT= 3                  'Assigns offset address
T$= "MTA UNL LISTEN 5"       'Assigns controller as the talker,
                             'all devices unlisten device 5
                             'listen
CALL TRANSMIT (T$, STATUS%)  'Transmits the command
```

LISTEN, TALK, and DATA commands are discussed next. To send data using a TRANSMIT routine, the interface must be a talker and one or more devices must be listeners. This is done by using the MTA (My Talk Address) command and the LISTEN command. SEND does both, the MTA and

LISTEN commands, and may be preferred in some applications.

LISTEN, TALK, and DATA are used with TRANSMIT and are the most commonly used commands. They are all followed by an argument list. In the case of LISTEN and TALK, the arguments are numbers that represent IEEE488 addresses. Arguments following a DATA command are single quoted strings or unquoted numbers that are interpreted as ASCII characters.

Any number of listeners can be specified in a single LISTEN command but there can be only one talker; only the last talk-address is significant.

In all cases, the arguments for LISTEN, TALK, and DATA terminate when the next command is found.

Examples

LISTEN 4 5 6	'Designates devices 4, 5, and 6 as 'listeners
TALK 2	'Designates device 2 as a talker
DATA '123.33'	'123.33 will be sent as ASCII data '(string)
DATA 13 10 34	'Unquoted numbers will be sent as binary 'data (in this case a carriage return, 'line feed, and double quote)
DATA 'IN; 123' 10	'Alphanumeric data followed by a line feed

Constants are transmitted as strings using the DATA statement. The maximum string length is set by the programming language and is 255 characters for interpreted BASIC and 32,767 characters for most compiled languages. Constants can be any printing or nonprinting character defined by the ASCII standard. By using various combinations of printing and non-printing characters in a DATA statement, it is possible to create any text sequence or control command.

Examples

DATA 'IN; DF; CN'	'Control statement with semicolon 'delimiters
DATA '3.141516, 5,'	'Numeric output with commas
DATA 'This is my name . . .' 03	'Text followed by an end of text (03) 'delimiter
DATA 'If I were a richman . . .' 13 10	'Text followed by a carriage return '(13) and a line feed (10)
DATA 13 10 'New line'	'Text preceded by a carriage return '(13) and a line feed (10)
DATA 27 '&k2S'	'A command string to a device 'preceded by an escape (27) 'character

Example. The following program lines transmit two lines of text (constants) to two printers simultaneously:

```
CONSTANT$= "MTA LISTEN 1 2 DATA 'alphanumeric' 13 10 'new line' "
CALL TRANSMIT (CONSTANT$, STATUS%)
```

MTA is required to establish the PC as a valid talker; LISTEN 1 2 tells printers located at addresses 1 and 2 that they are addressed to listen; DATA is the required keyword for the transmit routine in the interpreter; and 13 10 are the ASCII codes for carriage return and line feed.

Variables may be transmitted as single quoted strings in DATA statements. String variables can be entered without modification; however, numeric variables must be converted to strings. In BASIC this is done with the STR$ command. Numeric variables are automatically converted to strings when LPRINT is used to transmit variables.

Example

```
TRANSMIT$= "MTA DATA 'FREQ' " + STR$(X) + "?"
CALL TRANSMIT (TRANSMIT$, STATUS%)
```

Arrays can be transmitted using a FOR NEXT or WHILE WEND loop.

Example. Send the array A(I), which consists of 100 real numbers, to an instrument at address 14.

```
MDATA$= "MTA LISTEN 14 DATA"
FOR I= 1 TO 100
T$= MDATA$ + STR$(A(I)) + " ' "
CALL TRANSMIT (T$, STATUS%)
IF STATUS%<>0 THEN GOSUB 200
NEXT I
...
END
200 PRINT "NON-ZERO STATUS="; STATUS% : RETURN
```

RECEIVE. RECEIVE is to receive data from the IEEE488 and return the data to the calling routine in a string variable. The length of the string and status of the transfer are also returned. It can be called using

```
receive$= SPACE$ (MAX.LENGTH)
CALL offset (receive$, length%, status%)
```

- SPACE$ (MAX.LENGTH) assigns a string of spaces of length MAX.LENGTH to the received string.

- offset must be set to 6 for the interpreter BASIC and 33 for the compiler.
- receive$ is any alphanumeric variable; receive$ when returned after the call holds the data transmitted from the designated talker. RECEIVE returns when the assigned string is full, a line feed is received, or an EOI is received.
- length% is returned after the call and is equal to the length of the received string.
- status% is returned with one of the following values:

0, successful transfer.

2, tried to receive when not a listener. An MLA (My Listen Address) command must be used to define the interface as a listener.

8, bus timeout.

Example

```
DEF SEG = &HC000
TRANSMIT = 3
RECEIVE = 6
CMD$ = "MLA 1 TALK 12"
CALL TRANSMIT (CMD$, STATUS%)
R$ = SPACE$(40)
CALL RECEIVE (R$, LENGTH%, STATUS%)
```

If the received string is "12345,6789,0" then the variables will have the following values:

```
R$ = "12345,6789,0"
LENGTH% = 12
STATUS% = 0
```

Variables that are received can be processed in several ways by using BASIC string variable commands along with the LENGTH$ variable returned from the RECEIVE routine. Alphanumeric strings can be processed as follows:

```
R$ = SPACE$(255)
CALL RECEIVE (R$, LENGTH%, STATUS%)
IF STATUS <>0 THEN 1000                    'Process errors
R$ = LEFT$ (R$, LENGTH%)
```

where R$ contains the exact number of characters received in the transmission. Numeric variables are returned as character strings that may be preceded by an unknown number of spaces. In this case, the numeric value of the variable can be obtained using the VAL statement: X = VAL(R$). If the first character of R$ is not numeric, VAL(R$) will return 0. If two (or more) numeric values are returned delimited by a comma, or some other non-numeric character, the procedure shown below will extract the numeric values:

```
DEF FNCHOP$(X$)= RIGHT$(X$, LEN(X$)−INSTR(X$,","))
R$= SPACE%(255)
CALL RECEIVE(R$, LENGTH%, STATUS%)
X= VAL(R$)
Y= VAL(FNCHOP$(R$))
Z= VAL(FNCHOP$(FNCHOP$))
```

Any device that follows the IEEE Recommended Practice for Code and Format Conventions (IEEE Standard # 728-1982) will delimit data with a comma or a semicolon.

SEND. SEND establishes the PC as a talker, assigns a listener, and sends data followed by a null with EOI. It is called with

CALL offset (addr%, data$, status%)

- offset represents the offset into the current segment defined by the DEF SEG instruction. It must be set to 9 for the interpreter and 36 for the compiler.
- addr% is the address of the designated listeners.
- data$ is a data string.
- status% is returned after the call. Possible values are

  0, successful transfer.

  8, bus timeout.

Example. The following routine sends an alphanumeric string to a device at address 5:

```
DEF SEG= &HC000               'Assigns current segment
SEND= 9                       'Assigns the offset address
ADDR%= 5                      'Assigns the listen address
SEND$= "ALPHANUMERIC 123"     'String to be sent
CALL SEND (ADDR%, SEND$, STATUS%)
```

SPOLL. The SPOLL routine executes a serial poll for the given address and returns the poll byte received. Serial polling is a method of sequentially checking the status on a number of devices on the IEEE488. To do this, the service request (SRQ) line controlled by the device and the serial poll, performed by the controller, must work together. When SQR is asserted, the controller begins to sequentially gather status from each device. A feature of the serial poll is that the status information returned to the controller contains the service request status and information about the event that caused the service request. This single status byte is usually enough to determine the requirements of the device. This routine is called using

CALL offset (addr%, poll%, status%)

- offset represents the offset into the current segment defined by the DEF SEG instruction; it must be set to 12 and may be assigned any name (such as SPOLL).
- addr% is the address of the device to be polled.
- poll% is returned to the calling routine and is equal to the poll byte received.
- status% is returned after the call. A returned 0 means a successful serial poll, and 8 means a bus timeout.

Example

```
DEF SEG = &HC000          'Assigns current segment
SPOLL% = 12               'Assigns offset address
ADDR% = 4                 'Assigns serial poll address
CALL SPOLL (ADDR%, SPOLL%, STATUS%)
```

DEVICE. DEVICE is used to install one, two, or three IEEE488 device drivers in place of the default printer drivers. IEEE devices can then be accessed using the BASIC or system printer commands (LPRINT, LPRINT USING, PRINT#, PRINT#USING, LLIST, and COPY) or equivalent print statements in PASCAL, C, and FORTRAN. It is called with

CALL offset (addr1%, [addr2%], [addr3%], number of devices%)

- offset must be set at 18 and may be assigned any name (such as DEVICE).
- addr1% is the address of the device assigned to LPT1 or PRN. In contrast to addr2% and addr3%, this parameter is not optional.
- addr2% is the address of the device assigned to LPT2. It is optional.
- addr3% is the address of the device assigned to LPT3. It is optional.
- number of devices% is the number of devices installed. A minimum of one and a maximum of three are allowed.

Example 1. Install one printer at device address 1.

```
DEF SEG = &HC000
DEVICE = 18               'Assigns offset address
```

```
PRINTER%= 1                    'Assigns printer IEEE488 address
  N%= 1                        'to one device
CALL DEVICE (PRINTER%, N%)
                               'Installs the printer
SYSTEM                         'The new device is now recognized by the
                               'operating system
```

Example 2. Install printer (address 2), plotter (address 4), and digital voltmeter (address 7).

```
PRINTER%= 2
PLOTTER%= 4
DVM%= 7
NDEVICE%= 3
CALL DEVICE (PRINTER%, PLOTTER%, DVM%, NDEVICE%)
SYSTEM
```

ENTER. ENTER establishes the PC as a listener, assigns a talker, and then receives data from the talker. It is called with

```
receive$= SPACE$(MAX.LENGTH)
CALL offset (receive$, length%, addr%, status%)
```

- receive$ must be set to a string whose length is greater than or equal to the length of the received string. The maximum length of a received string is 255 characters.
- SPACE$(MAX.LENGTH) assigns a string of space of length MAX.LENGTH to the received string.
- offset must be set at 21 for the BASIC interpreter and at 39 for the compiler; it must be assigned any name (such as ENTER).
- receive$, when returned after the call, holds the transmitted string from the designated talker.
- length% is returned after the call and is equal to the length of the received string.
- addr% is the address of the designated talker.
- in a successful transfer status% returns 0.

ENTER returns when the assigned string is full, a line feed is received, or an EOI is received.

Example

```
DEF SEG= &HC000
ENTER= 21
R$= SPACE$(80)
```

DEVICE% = 6
CALL ENTER (R$, LENGTH%, DEVICE%, STATUS%)

Single Line Commands. Single line commands are used to control 4 of the 5 control lines: Interface Clear (IFC), Remote Enable (REN), Attention (ATN), and End or Identify (EOI). As previously stated, each command has a single bus line dedicated for its use. IFC and REN are only available to the system controller. ATN can be used by any controller and EOI by any device.

IFC can be asserted by using the IFC command in a TRANSMIT statement or by calling INITIALIZE with a level 0.

Example 1

IFC$ = "IFC"
CALL TRANSMIT(IFC$, STATUS%)

Example 2

INITIALIZE = 0
MY.ADDRESS = address of interface
SYSTEM.CONTROLLER% = 0
CALL INITIALIZE (MY.ADDRESS, SYSTEM.CONTROLLER%)

REN is asserted by calling TRANSMIT with a string argument of REN.

Example

REN$ = 'REN'
CALL TRANSMIT(REN$, STATUS%)

ATN is asserted whenever a valid multiline, address, or secondary command is sent by using TRANSMIT.

EOI is asserted by use of the EOI command.

Example

T$ = "MTA DATA 'AB' EOI 'C' "   'C is sent with EOI on
CALL TRANSMIT (T$, STATUS%)

Multiline Commands. Multiline commands can be sent using the TRANSMIT routine. These mnemonics cause specific messages to be placed on the data bus while ATN is asserted. These commands are called multiline, because they use the data lines to transfer the command, rather than a single dedicated bus-management line. The results of the command

and the action taken by a device are unique to every device and system. These commands are defined as follows:

CMD (command)	Allows any numeric or string expression to be sent with ATN true. Numbers must be integers that are rounded modulo 256 and sent as one byte.
DCL (device clear)	A device will be initialized to its power-on status.
LLO (local lockout)	This command causes the device to ignore its front panel controls.
SPE (serial poll enable)	This command does not perform a serial poll; it only enables a device to place its serial poll response on the IEEE488.
SPD (serial poll disable)	It disables the serial poll response of a device.
PPU (parallel poll unconfigure)	This command disables the parallel poll response of all devices.
UNL (unlisten)	On receipt of this command all listeners will stop listening.
UNT (untalk)	On receipt of this command the current talker will stop talking.

Example. Instruct all devices to stop talking, stop listening, and perform a device clear.

```
COMMAND$ = "UNT UNL DCL"
CALL TRANSMIT(COMMAND$, STATUS%)
```

Addressed commands. There are five addressed commands:

GET (group execute trigger)	Used to simultaneously trigger a group of devices.
SDC (selected device clear)	On receipt of this command a device currently addressed to listen is initialized to its power-on state. SDC clears only the active listeners.
GTL (go to local)	On receipt of this command a device resumes control from its front panel.
PPC (parallel poll configure)	Allows the controller to remotely program the parallel poll response to a specific device.

TCT (take control)   Allows the current controller to pass
                                        control to a new controller.

To execute these commands, the TRANSMIT routine and various control commands can be used. To execute a command, the command string is set equal to the control string shown below and then TRANSMIT is called:

GET$ = "MEA UNL LISTEN listen addresses GET"
SDC$ = "MTA UNL LISTEN listen addresses SDC"
GTL$ = "MTA UNL LISTEN listen addresses GTL"
PPC$ = "MTA UNL LISTEN listen addresses PPC CMD ppoll#"
TCT$ = "TALK talk address TCT"

Example. Simultaneous triggering of a group of devices.

DEF SEG = &HC000
TRANSMIT = 3
GET$ = "MTA UNL LISTEN 6 GET"
CALL TRANSMIT(GET$, STATUS%)   'Executes a group execute
                                        'trigger

As application examples, two general purpose, useful programs are presented next. Both are taken from the PC<=>488 Programming and Reference Manual from CEC, with permission. The second program has been adapted by us for the continuous acquisition of data through a Keithley multitester connected to an IBM PC. Note that the address of the TRANSMIT and RECEIVE routines are not the same in Examples 1 and 2, since the former is written for interpreter execution whereas the latter must be compiled.

Example 1. Program for ASCII file transfers between two personal computers via the IEEE488 interface bus.

```
100 REM  Run this program then send data from the
110 REM  source computer to device address one.
120 REM      The IBM PC appears as a printer at address one.
130 REM  To save the transferred file on disk, provide a
140 REM      filename when prompted.
150 REM  --------------------------------------------------------------------------------
160 CLS
170 DEFINE A-Z
180 LINE.COUNT = 0
190 PRINT   "NOTE: you must set your PC-488 to act as a device"
200 PRINT   "(not as a system controller)"
210 PRINT   "---Hit RETURN to continue"
220 PRINT   "--------------------------------------------------------------------------------"
```

```
230 LINE INPUT R$
240 DEF SEG= &HC000                    'Firmware segment address
250 INIT= 0: XMIT= 3: RECV= 6          'Firmware offset address
260 R$= SPACE$(80)                     'Receives string length
270 PC.ADDRESS= 1 : DEVICE= 2          'PC IEEE488 address 1 as device
280 INPUT "Enter file name to save or <return> to end", FILENAME$
290 IF FILENAME$= " " THEN STOP
300 OPEN FILENAME$ FOR OUTPUT AS #1
310 CALL INIT (PC.ADDRESS; DEVICE)
                                       'IBM PC initialized as device
320 IF INP(&H2BC) AND 4 THEN 330 ELSE 320
                                       'If addressed then receive
330 CALL RECV(R$, L, S)
340 R$= LEFT$(R$, L)                   'Adjusts the string to length
350 PRINT R$                           'Displays transferred line
360 PRINT #1, R$                       'Saves line to file
370 LINE.COUNT= LINE.COUNT+1           'Increases line counter
380 R$= SPACE$(80)
390 IF S<>8 THEN 330                   'Continues until timeout
400 REM
410 CLOSE
420 PRINT LINE.COUNT; "lines transferred"
                                       'Displays number of transferred lines
430 END
```

Example 2. BASIC program to acquire data from a Keithley multitester through an IEEE488 interface. The multitester must be provided with this optional interface, and the PC< = >488 interface must be plugged in one of the slots of the PC. The program must be compiled before execution.

```
DEF SEG= &HC000: RECALL.PUT= 1: RECALL.COUNT= 0
S$= SPACE$(80)                         'Used to erase an input line
RECALL.MAX= 6                          'Controls the depth of the RECALL
                                       'BUFFER
DIM RECALL.BUF$ (RECALL.MAX)
BLANK$= "                                                           "
INITIALIZE%= 0: TRANSMIT%= 30: RECEIVE%= 33: MY.ADDRESS%= 21
ST$= "UNL UNT MLA 1 TALK 24"
SYSTEM.CONTROLLER%= 0
    CALL ABSOLUTE (MY.ADDRESS%, SYSTEM.CONTROLLER%,
        INITIALIZE%)
    PRINT "Press any key to continue": XLINE= CSRLIN
    WHILE INKEY$= " ": WEND
    n= 0: m= 0: LOCATE XLINE − 1
250 'Command entry point
    IF INKEY$= "E" OR INKEY$= "e" THEN END
    If n= 15 THEN
```

```
      n = 1: LOCATE 8, 1: PRINT BLANK$: PRINT BLANK$:
      LOCATE 8, 1
   END IF
   IF C$ = "T" THEN C$ = "R" ELSE C$ = "T"
   IF C$ = "R" THEN 320
   IF C$ = "T" THEN 440
320 'Receive utility
   R$ = SPACE$(30)
   CALL ABSOLUTE (R$, LENGTH%, STATUS%, RECEIVE%)
   R = VAL (MID$(R$, 5))
   IF m = 1 THEN
      XLINE = CSRLIN: PRINT BLANK: PRINT BLANK: LOCATE
      XLINE
   END IF
   PRINT MID$ (R$, 1, 4); R, : m = m + 1
   IF m = 6 THEN m = 1: n = n + 1
GOTO 250                          'Returns to command entry
440 'Transmit utility
   RECALL.BUF$ (RECALL.PUT) = ST$
   RECALL.GET = RECALL.PUT: RECALL.PUT = RECALL.PUT + 1
   IF RECALL.COUNT < RECALL.MAX THEN
      RECALL.COUNT = RECALL.COUNT + 1
      ELSE RECALL.COUNT = RECALL.MAX
   END IF
   IF RECALL.PUT > RECALL.MAX THEN RECALL.PUT = 1
   CALL ABSOLUTE (ST$, STATUS%, TRANSMIT%)
GOTO 250                          'Returns to command entry
```

CHAPTER

7

INSTRUMENTS AND APPARATUS FOR LABORATORY AUTOMATION

7.1. INTRODUCTION

Today many scientific instruments are provided with the appropriate interfaces for communication, whether in its standard configuration or as an extension option. In some cases, the interface is designed only to output data. Other instruments are capable of higher communication levels and, in the best cases, it is possible to control the whole instrument through an external standard communication interface. Communication capability is an important feature to be considered before purchasing any instrument.

With the aim of introducing the reader to the methodology and problems related with constructing modular automated systems, several apparatus and laboratory instruments prepared for digital communications are described in this chapter. This is not an exhaustive description but rather a collection of some basic examples. The instruments described here are necessarily from a limited number of manufacturers; however, many other manufacturers offer similar instruments with the same or better performances.

With some knowledge of analog and digital electronics, instruments that are not prepared for communication can also be used in automatic systems. Once transduced and conditioned, all instruments present the signal in a readable form. A simple traditional way is the displacement of a needle in an analog display. Digital presentations are more convenient and most frequently found in modern instruments.

In the first case, the signal can easily be acquired by means of an A/D converter. Many instruments also have an analog output that is characterized by a fixed voltage range for connection to a paper chart recorder. This output can also be used to feed an A/D converter, but usually a voltage adapter or programming of the A/D converter input gain is required.

A good solution is to use a digital multitester to acquire the signal. Quality multitesters measure several electrical parameters (voltages, currents, resistances, etc.). They may be provided with automatic selection of the most convenient scale to acquire the input signal; they then convert the signal to digital and provide a standard output, usually an IEEE488.

Unfortunately for high-rate applications, most low-cost testers have a low reading rate. The maximum working frequency is usually limited by the display refreshment rate, which is on the order of a few times per second. Instead, any common A/D converter has an input rate of, at least, a few hundred readings/s. Also, in comparison with A/D converters that usually provide a multiplexed input, multitesters have a single input channel.

Many modern instruments internally convert the signal from analog to digital and provide only a digital output. In this case, data acquisition with a computer may present some difficulty. Sometimes, the instrument will be provided with a BCD interface for connection of a printer, which may be used for communication with the computer. Communication can be very easily implemented when the instrument is provided with a standard interface, such as the parallel IEEE488 or the serial RS232C. Again, a limitation of the latter is its low output rate, frequently no more than a few data per second.

7.2. THE KEITHLEY 175 MULTIMETER

The Keithley Model 175 is a digital multimeter that performs voltage, current, resistance, and dB (decibel) measurements with autoranging capabilities. The decibel function makes it possible to compress a large range of readings into a much smaller scope. Some specifications are summarized in Table 7.1.

Ranging may be shifted to automatic on the voltage and resistance scales with an autoranging time of 300 ms per range. It has a REL (relative measures) option to give readings with respect to any base line. Also, data logger operation is covered by the tester with a storage capacity of 100 readings. It can detect and store maximum and minimum readings continuously while in the data logger mode. Data are recorded at one of six selectable rates from 3 readings/s to 1 reading/hour. Maximum conversion rate is 3 readings/s.

The capabilities of Model 175 can be enhanced with the addition of the IEEE488 optional interface Model 1753, designed to interface it to the IEEE488 bus. The volt (ac and dc) and ohm ranges and the dB and REL features are controlled by programming commands over a IEEE488-1978 standard bus. The specifications of the interface are summarized in Table. 7.2.

The base address of the IEEE488 interface may easily be set by using five address switches on the rear panel of Model 175. The base address of the Model 1753 interface is set at 24 at the factory.

Table 7.1. Specifications of the Keithley Model 175 Multimeter

dc Voltage Ranges[a]		TRMS ac Voltage Ranges[b]	Resistance Ranges	
200 mV	± 10 μV	200 mV	200 Ω	± 10 mΩ
2 V	± 100 μV	2 – 750 V	2 kΩ	± 100 mΩ[c]
20 V	± 1 mV		20 kΩ	± 1 Ω
200 V	± 10 mV		200 kΩ	± 10 Ω[c]
1000 V	± 100 mV		2 MΩ	± 100 Ω
			20 MΩ	± 1 kΩ
			200 MΩ	± 100 kΩ

dc Current Ranges		TRMS ac Current Ranges
200 μA	± 10 nA	200 μA – 20 mA
2 mA	± 100 nA	200 mA
20 mA	± 1 μA	2000 mA
200 mA	± 10 μA	10 A
2000 mA	± 100 μA	
10 A	± 1 mA	

[a] Input resistance: 11 MΩ on all scales, except 10 MΩ on 200 V and 1000 V ranges.
[b] Because of its TRMS (True Root Mean Square) measuring capabilities, Model 175 provides accurate ac measurements for a wide variety of ac input waveforms.
[c] The 2 kΩ and 200 kΩ ranges can be used for testing semiconductor junctions by pressing the 2 kΩ and 200 kΩ push-buttons simultaneously.

Table 7.2. Model 1753 IEEE488 Interface Specifications

Feature	Specification
Trigger to first byte out	700 ms except 1 s on ohm in T0 and T1 modes
	800 ms except 2 s on ohm in T3 mode
Address modes	Talk Only Mode and Addressable Mode
IEEE488 bus implementation	Multiline commands: DCL, SDC, GET, GTL, UNT, UNL, SPE, SPD
	Unline commands: IFC, REN, EOI, SQR, ATN
	Interface functions: SH1, AH1, T5, TE0, L4, LE0, SR1, RL2, . . .
	Programmable parameters: Range, REL, dB, EOI, Trigger, Calibration, SRQ, Status, Output, Format, Terminator

The interface may be set for one of two operation modes. In the talk mode, Model 175 outputs data to other devices (e.g., a printer). In the addressable mode, Model 175 can both receive commands and transmit data over the bus through the interface. Model 1753 must be set to the addressable mode when used with an external controller. Mode selection is done with the TO/ADDRESSABLE mode selection switch located on the rear panel of Model 175.

Once the device is addressed to talk or listen, the appropriate bus transactions will take place. For example, if Model 1753 is properly addressed to talk, it will normally place its data string on the bus one byte at a time. The controller will then read this information, and the appropriate software can then be used to channel the information to the desired location.

Model 175 data are transmitted over the bus as a string of ASCII characters with the format shown in Figure 7.1. The mantissa of the reading is made up of seven characters, including sign and decimal point, while the exponent requires three characters.

To obtain the data string from the instrument, the controller must perform the following sequence:

1. Set ATN low.
2. Address the instrument to talk.
3. Set ATN high.
4. Input the data string one byte at a time.

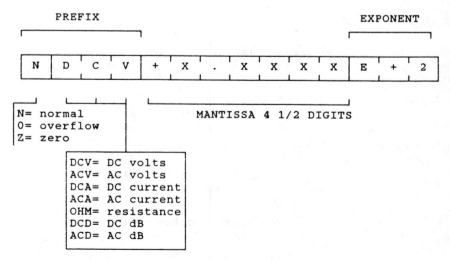

Figure 7.1. Format of the data transmitted by the Model 175 multimeter.

Programming of Model 175 is done by sending the appropriate device-dependent commands through the interface. These commands are given in Table 7.3.

Access to information concerning present operating modes of the instrument is allowed by a series of status word commands. When a status word command is given, Model 175 will transmit status information instead of its normal data string the next time it is addressed to talk. The status byte will contain information relating to data and error condition within the instrument. When a particular bit is set, certain conditions are present. Figure 7.2 shows the general format for the status word commands.

An example of a program written in BASIC which acquires data from Model 175 is given below. Data are acquired continuously, until key E (end) is pressed. Contrary to the program given at the end of Chapter 6 (Section 6.3.7), the present program must be executed in interpreted BASIC. Note that the syntax of the instruction CALL and the address offsets are not the same. This program was checked using the IEEE488 interface from CEC (Capital Equipment Corporation), model PC<=>488, installed in a slot of an IBM PC.

```
100     CLS
110 'Initialization
120     FOR I = 1 TO 10: PRINT : NEXT
130     DEF SEG= &HC000
140        RECALL.MAX= 6                  'This controls the depth of the
                                          'RECALL BUFFER
150        DIM RECALL.BUF$(RECALL.MAX)
160        RECALL.PUT= 1
170        RECALL.COUNT= 0
180        S$= SPACE$(80)                 'Used to erase an input line
190        INITIALIZE= 0
200        TRANSMIT= 3
210        RECEIVE= 6
220        MY.ADDRESS%= 21
230        SYSTEM.CONTROLLER%= 0
240     CALL INITIALIZE (MY.ADDRESS%, SYSTEM.CONTROLLER%)
250        FOR I= 1 TO 50: NEXT
260     IF INKEY$= "e" OR INKEY$= "E" THEN END  ...
                                          'End of process
270 'Command entry point
280     N= 0
290     C$= " "
300     IF C$= "T" THEN C$= "R" ELSE C$= "T"
310     IF C$= "C" THEN CLS : GOTO 270
320     IF C$= "R" THEN GOTO receive
```

Table. 7.3. Summary of the Device-Dependent Commands of the Keithley Model 175 Multimeter

Type	Command	Description	
dB	D0	dB off	
	D1	dB on	
Range		Volts	Ohms
	R0	autorange	autorange
	R1	200 mV	200 Ω
	R2	2 V	2 kΩ
	R3	20 V	20 kΩ
	R4	200 V	200 kΩ
	R5	2000 V	2 MΩ, 20 MΩ, 200 MΩ
Relative	Z0	REL off	
	Z1	REL on	
	T0	Continuous talk	
	T1	One-shot on talk	
	T2	Continuous on GET	
	T3	One-shot on GET	
	T4	Continuous on X	
	T5	One-shot on X	
Execute	X	Execute other device-dependent commands	
EOI	K0	EOI enabled	
	K1	EOI disabled	
Status word	U0	Output status word	
Data	G0	Readings and status word with 175 prefix format	
	G1	Readings and status word without 175 prefix format	
SRQ	Mnn	SRQ on error and/or data conditions	
Store	L0	Store calibration constants	
Terminator	Y(ASCII)	ASCII character	
	Y(LF)	CR LF	
	Y(CR)	CR	
	Y(DEL)	None	
Digital calibration	V+n.nnnnE+nn	n represents calibration value	

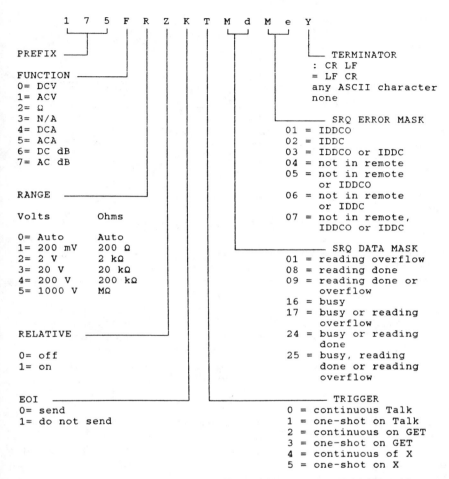

Figure 7.2. Format for the status word commands of the Keithley Model 175 multimeter.

```
330    IF C$= "T" THEN GOTO transmit
340       GOTO 270                        'If invalid entry
350 'Receive utility
360          R$= SPACE$(30)
370    CALL RECEIVE (R$, LENGTH%, STATUS%)
380          R= VAL (MID$(R$, 5))
390          PRINT MID$ (R$, 1, 4); R,
400    GOTO 270                           'Return to command entry
```

```
410 'Transmit utility
420        ST$= "UNL UNT MLA 1 TALK 24"
430        RECALL.BUF$ (RECALL.PUT)= ST$
440        RECALL.GET= RECALL.PUT
450        RECALL.PUT= RECALL.PUT + 1
460        IF RECALL.COUNT < RECALL.MAX THEN
           RECALL.COUNT= RECALL.COUNT + 1 ELSE
           RECALL.COUNT= RECALL.MAX
470        IF RECALL.PUT > RECALL.MAX THEN RECALL.PUT= 1
480    CALL TRANSMIT (ST$, STATUS%)
490    GOTO 300                      'Return to command entry
```

7.3. POTENTIOMETERS

A potentiometer is a voltmeter with a high input impedance, which is required to measure the signals produced by low output impedance transducers. pH meters are potentiometers whose output is given as a pH scale. Potentiometers usually have a recorder analog output, and many modern quality models also have a standard communication interface. The latter may be used to perform two very important functions in laboratory automation: A/D conversion and communication interface. This is accomplished by connecting the analog signal, which must be measured, to the electrode analog input of the potentiometer. A voltage divider or an amplifier may be necessary to take advantage of the whole scale range. Data acquisition rate is usually below 3 readings/s, which is enough in many laboratory applications.

As an application example let us see how the Crison Digilab 501 (± 1 mV precision) and 517 (± 0.1 mV precision) potentiometers/pH meters communicate. These instruments are provided with an RS232C interface with the following specifications: 2400 baud, 1 start bit, 7 bit word length, 1 parity bit, 2 stop bits, and even parity control.

To achieve compatibility, the same communication protocol was adopted

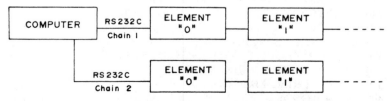

Figure 7.3. A computer supporting two serial communication chains.

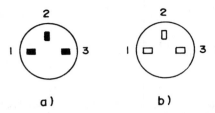

Figure 7.4. Scheme of the connectors of the Crison instruments for serial communication. From pin 1 to 3: talker, common ground, and listener lines; (a) to the controller (female) and (b) to the next element (male).

for a series of Crison instruments, such as the burettes of Model 738. This permits connection of several instruments to the same RS232C port of the computer using a linear chain topology. When the instruments do not use the same protocol, additional RS232C interfaces are required. The system configuration is schematized in Figure 7.3.

The connectors of the potentiometers are represented in Figure 7.4. Since pins 1 and 3 are equivalent to pins 2 and 3 of the RS232C interface, lines 1 and 3 should cross each other between Crison instruments when connected in a chain fashion (Figure 7.5).

The working protocol (channel opening, element numbering, etc.) has been described in Chapter 6. When requested for a reading, the potentiometer responds with a code, which may contain the following:

"+" or "−"	Sign of the reading
XXXXX	0 to 9 digits which correspond to the values of pH or mV, without decimal point
"0" or "H"	Code of function (mV or pH)
"?"	Receipt of an unrecognized command

To request a reading, the computer sends the identification number of the element. Thus, for a potentiometer located in the first place of a chain, which communicates through channel #2, a "0" is sent:

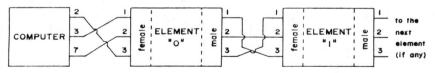

Figure 7.5. Connection of compatible Crison instruments in a serial chain.

PRINT #2, "0"

The response of the potentiometer is

N/LF/SXXXXXF/CR/LF

where the slashes have been added for clarity, and where N is the identification number, LF is the line feed, S is the sign, the Xs are digits representing the measure, F is the code of the signal (0 for mV and H for pH), and CR is carriage return. Thus, for the value −512 mV, the replay of the potentiometer located in the "0" position of the chain is

```
0
(blank line)
−005120
```

The design and characteristics of the Crison potentiometers/pH meters Models 2001 (± 1 mV precision) and 2002 (± 0.1 mV precision) are entirely different. They are provided with a single RS232C interface, and therefore they can only be connected at the end of serial communication chains, as the last element of the chain. Their communication protocol is 1200 baud, 1 start bit, 8 bit word length, no parity bits, and 2 stop bits. This is not the same protocol used by the Crison 501 and 517 potentiometers and 738 burettes, and therefore a different RS232C interface must be used to connect them.

The Crison 2001 and 2002 potentiometers do not output information upon request. Instead, information is sent in an independent continuous fashion, each time the display is refreshed (each $\frac{1}{3}$ s). Therefore, only two of the three pins of the communication connector are really used: pin 1 for data output and pin 2 as ground. The data have the following format (slashes added):

V:/SP/SP/readout/SP/U/CR/LF/T:/SP/SP/readout/SP/°C/CR/LF/LF

where the characters V: and T: are used to synchronize the readings with the computer. The computer must search for the characters V: to acquire a voltage, and the same must be done with the following temperature reading. Readout formats are XXXX.X (mV) or XX.XXX (pH or pX) and XXX.X (temperature in °C). Other symbols are SP, space; U, units used (mV, pH or pX); all other characters are ASCII codes.

7.4. SPECTROPHOTOMETERS

The simplest way of implementing data acquisition of absorbances is the connection of the recorder analog output of the spectrophotometer to an A/D converter. This system is very adequate when working at a single wavelength, but some problems will probably appear in the acquisition of spectral data, particularly when filter or lamp changes occur.

Many modern spectrophotometers are provided with devoted microprocessors for control of the monochromators, data treatment, and application of specific procedures. Some of them have optional standard interfaces for communication and a few also permit external control of the instrument from a computer. As an excellent example of the latter, the Hewlett-Packard 8452A diode array UV–visible spectrophotometer is described next. This instrument is particularly adequate when used in automatic modular systems such as a spectrophotometer, a dedicated detector, or a specialized analyzer.

The diode array technology allows multiwavelength measurement almost simultaneously and greatly enhances the quantitation capability and speed, which permits many new applications. Data can be acquired at very high rates, every 2 nm from 190 to 820 nm. The spectrophotometer specifications are given in Table 7.4.

The instrument does not have a front panel for manual control; it can only be controlled through an IEEE488 or an HP-IL interface. The instru-

Table 7.4. Specifications of the HP 8452A Spectrophotometer

Feature	Values
Full spectrum scan	0.1 s
Data repeatedly acquired	
Up to 35 wavelengths	0.1 s
Full spectrum scan	0.6 s
Wavelength range	190–820 nm
Wavelength accuracy	± 2 nm
Wavelength reproducibility	± 0.05 nm
Spectral bandwidth	2 nm
Photometric range	0.0022–3.3 AU
Photometric accuracy	± 0.005 AU
Noise	<0.0002 AU rms[a]

[a]rms = root mean square.

ment is shipped with the default IEEE488 address 18. A general purpose 8-bit input port and an 8-bit output port are built in and available through a 37-pin connector. These ports are programmed by the controller to manipulate accessories such as a peristaltic pump and an autosampler. One of the input bits can be programmed to trigger spectral scans in the instrument or to interrupt the controller.

The instrument has three communication buffers, which can be monitored through the STATUS commands. The *input buffer* is 256 bytes long and is used to receive ASCII instructions. The *measurement data buffer* is used to store measurement data as a result of the MES or sometimes REF instructions. This buffer is 6588 bytes long and may contain up to four full binary spectra or one full ASCII spectrum. The *reply buffer* contains output data other than the measurement data. This buffer is 256 bytes long. All the replies to instructions such as status and parameter inquiries are stored in this buffer.

Communication and control of the instrument can be achieved by simple input/output statements in a high-level language. In the following example, written in HP BASIC, absorbances and variances are obtained at 300 nm:

```
REAL Absorbance, Variance        ! Data variables
OUTPUT @Instr; "DFL"             ! Set default parameters
OUTPUT @Instr; "WAV 1,300"       ! Set wavelength to 300 nm
OUTPUT @Instr; "FMT 0,1"         ! Ask data in ASCII
OUTPUT @Instr; "REF 0.5"         ! Take blank, 0.5 s integration time
OUTPUT @Instr; "MES"             ! Take a scan
ENTER @Instr; Absorbance, Variance  ! Get data
```

Hewlett-Packard provides a program package containing a series of basic procedures such as an instrument autocheck, recording of spectra, application of basic quantitative procedures, and recording of measurements at several wavelength values simultaneously as time functions. This is a very useful package, but full advantage of the instrument's capabilities can only be taken if the user develops new programs, designed to meet specific requirements.

For this purpose, the instruction set given in Table 7.5 must be used. Use of the instructions must take into account the commands of the IEEE488 interface. The supporting software library provided with the interface may also be used. This software, which can be compiled with the QuickBASIC compiler, makes it easier to use the commands of the instrument.

The user can also contact Hewlett-Packard to obtain uncompiled programs in BASIC, which obviously would be very helpful in developing the user's own programs. As an example, a BASIC subroutine for autochecking of the instrument is given next.

Table 7.5. Instruction Set of the HP 8452A Diode Array Spectrophotometer

Instruction	Description	Type[a]
ABT	Abort, return to a known state	[C]
BUR (burst mode)	Burst, changes process priority	[I]
CAL	Read current wavelength calibration	[O]
CCK	Clear frame clock	[C]
CEL (cell number)	Multicell transport cell number	[C]
CLK	System clock	[O]
CMS	Current measure status	[O]
DFL	Set default instrument parameters	[C]
DRS	Data record size	[O]
ERR	Read last error encountered	[O]
FMT (record length mode), (format)	Format	[I]
GAJ	Gain adjust of the A/D converter	[C]
HOM	Multicell transport home	[C]
IDY	Identify instrument and firmware	[O]
INT (mode), (gain)	Intensity	[I]
LMP	Control power to the lamp	[C]
LPS	Lamp status	[O]
LST (parameter number)	List parameters	[O]
MEM (address)	List memory (diagnostics)	[O]
MES	Start measurement cycle	[C]
MSK (mask)	Service request mask	[I]
POK (address), (value)	Poke memory	[I]
RCK	Read frame clock	[O]
RDP (port)	Read from port GPIO	[O]
REF (integration), [(output)]	Reference spectrum	[C]
SHU (mode)	Shutter control	[I]
STA	Status (RS232C)	[O]
STP (step number)	Multicell step number	[C]
TIM (integ), (interval), (reading), (delay)	Time parameters	[I]
TMS	Test measure (dark current)	[C]
TRA	Trigger acknowledge	[C]
TRE (data type), [(error band)]	Test reference (analog electronics)	[C]
TRG (state), (means flag), (lag time)	Trigger (synchronism)	[I]
TRS (reply type)	Diagnostic test results	[O]
TST (test number)	Diagnostic test (hardware)	[C]
VRM (mode)	Variance	[I]
WAI	Wait	[C]
WAV (mode), (wl 1), (wl 2), . . . , (wl 20)	Wavelength	[I]
WRP (port), (value)	Write to port	[C]

[a] [I] are instructions that contain input data to the instrument. [O] are instructions requesting output data from the instrument. [C] are instructions that trigger some type of instrument or device control.

239

```
CMD$= "TST 0"                        'Diagnostic test of hardware
LENGTH= LEN(CMD$)
CALL IOOUTPUTS (HP8452, CMD$, LENGTH)
                                     'Request perform diagnostic test
GOSUB errors                         'Goes to the errors subroutine check
.........................................
CMD$= "TRS 0"                        'Diagnostic test results
LENGTH= LEN(CMD$)
CALL IOOUTPUTS (HP8452, CMD$, LENGTH)
                                     'Request diagnostic results
GOSUB errors                         'Goes to the errors subroutine check
.........................................
RES$= SPACE$(10)
MAX.LENGTH= 10
CALL IOENTER (HP8452, RES$, MAX.LENGTH, ACT.LEN)
                                     'Inputs results
.........................................
```

7.5. SPECTROFLUORIMETERS

Data acquisition and control in fluorimetry making use of the spectrofluorimeter LS-5 from Perkin-Elmer is described here. This instrument has an optional RS232C interface that allows easy data acquisition and control of all operating parameters via a user friendly set of commands.

The new Perkin-Elmer Model LS 50 looks indeed very different from Model LS-5, since it is designed for computer control and has no front panel for manual control. However, with the exception of some instrumental improvements, the basic set of commands and the way control can be taken by an external computer are the same as those of Model LS-5.

The LS-5 *communication protocol* is: selectable rate for 300, 600, 1200, and 2400 baud through an on-board switch; full duplex (simultaneous bidirectional transmission capability); 1 start bit; 7 bit word length; 1 stop bit; even parity; no terminating sequence (carriage return or line feed will prevent communication). Thus, for instance, the opening of the I/O ports on an IBM-PC may be

    OPEN "COM1:2400,E,7,1" AS #1

All commands would be issued as print statements to the spectrofluorimeter:

    PRINT #1, "$FS X"

The RS232C *cable configuration* contains pins 2 to 8 and 20 with their usual meaning:

Pin 2   Tx, transmitted data
Pin 3   Rx, received data
Pin 4   RQS, request to send
Pin 5   CLS, clear to send
Pin 6   DSR, data set ready
Pin 7   GRD, signal ground
Pin 8   RCD, received carrier detect
Pin 20   DTR, data terminal ready

Table 7.6. Synopsis of Perkin-Elmer LS-5 Spectrofluorimeter Commands via the RS232C Interface

Command	Description
$AC I	Sets autoconcentration mode
$AH I	Sets abscissa high limit
$AL I	Sets abscissa low limit
$AZ I	Selects auto zero mode
$CH X	Selects move on chart paper (in mm)
$CY X	Sets cycle time
$EC S	Diagnostic command to echo back string
$ET	Returns current cycle time and minimum cycle time
$FL I	Selects fluorescence mode
$FR X	Selects filter response
$FS X	Sets fixed scale
$GM I	Selects GOTO emission wavelength
$GX I	Selects GOTO excitation wavelength
$IP I	Sets integrated phosphorescence mode
$LC I	Selects liquid chromatography (LC) mode
$LP I	Selects LC mode, but requires no time entry
$MA	Selects manual mode
$MX I	Selects monochromator
$NR	Selects normal scan
$PB I	Replots binary data on recorder
$PD I	Sets phosphorescence delay
$PG I	Sets phosphorescence gate
$Ph I	Selects fluorescence or phosphorescence mode
$PS	Selects pre-scan mode
$RD	Performs an integration and returns an ordinate value
$RE I	Selects free-running or unprompted mode
$RS X	Sets recorder format
$SC I	Selects scan mode
$SS X	Selects scan speed
$ST	Returns status of instrument
$TD I	Selects time drive mode
$TI X	Sets time interval for time drive

The *command format* is a two-letter code, always preceded by a $. If the control requires a numeric parameter, then it follows the command format preceded by a space. Table 7.6 presents a synopsis of the most important commands.

When the argument is an I, the number entered should be an integer. An X argument requires a real number. The monochromator slits, shutter, or sample position cannot be controlled by the computer.

The RS232C interface can be set to operate in a prompted or a free-running mode. The prompted mode requires that, after a command is sent to the instrument from the computer, a prompting character be sent to elicit a response. After the LS-5 receives this prompt, it responds by first sending a four-number handshake code, the meaning of which is defined in Table 7.7. Let us examine two examples of how data can be acquired and the instrument controlled.

Example 1. For a discrete fluorescence reading the following sequence of QuickBASIC instructions may be used:

```
OPEN "COM1:2400,E,7,1" AS #1        'Opens communication port
PRINT #1, "$RE 0" : INPUT #1, d$     'Selects free-running mode
'Select wavelengths
```

Table 7.7. Handshake Response Codes of the LS-5

Code	Meaning
0000	No error
0001	Command conflict
0002	Verb not recognized or $ not present
0003	Unexpected parameter
0005	No parameter
0006	Illegal character in numeric field
0007	Parameter too large to handle
0008	Negative parameter not acceptable
0009	Value out of range for this command
0010	Parameter conflict between commands
0013	Pre-scan functional error
0014	Input buffer overflow
0015	Wavelength of LC-scan error
0016	Instrument busy in manual
0017	Cycle time too short for repeat scans
0022	Character parity, framing, or overrun error
0023	Output buffer overflow
0099	Abort requested

```
INPUT "Enter excitation wavelength= "; EX
INPUT "Enter emission wavelength= "; EM
EX$= "$GX" + STR$(EX)
EM$= "$GM" + STR$(EM)
PRINT #1, EX$: INPUT #1, D$        'Sends monochromators to
                                   'appropriate wavelengths
PRINT #1, EM$: INPUT #1, D$
PRINT #1, "$FR 0": INPUT #1, D$    'Sets integration time
'Readings
PRINT #1, "$RD": INPUT #1, D$      'Sends read command
INPUT #1, NUM$: INPUT #1, D$       'Inputs results
'Begin data format translation
FLG$= LEFT$ (NUM$, 1)
IF FLG$= "−" THEN GOTO negative ELSE positive
negative:
X$= MID$ (NUM$, 2, 1)             'Decimal point position
X1$= RIGHT$ (NUM$, 3)             'String of the digits
X1$= FLG$ + X1$                   'String of the sign + digits
GOTO numeric
positive:
X$= MID$ (NUM$, 1, 1)             'String of the decimal point position
X1$= RIGHT$ (NUM$, 4)            'String of the digits
numeric:
X= VAL(X$)                        'Value of the decimal point position
X1= VAL(X1$)                      'Value of the digits
X= X1/10^X                        'Value with correct decimal point
                                  'position
```

In this example, a read command ($RD) performs an integration for a time that depends on the response number selected and returns the ordinate reading on the display. The format of the number sent back is XYYYY or −XYYY, where the Ys are the value and X is the position of the decimal point, beginning at 0, which represents the extreme right position.

Example 2. To perform a scan, the sequence may be:

```
OPEN "COM1:2400,E,7,1" AS #1            'Opens communication ports
INPUT "Excitation wavelength= "; EX
INPUT "Emission wavelength= "; EM
INPUT "Emission ending wavelength= "; EMEND
                                        'Data taken every 0.5 nm
NUMBERPTS= (EMEND-EM)/0.5+1             'Number of data points
DIM VALUE(NUMBERPTS)
PRINT #1, "$RE 0": INPUT D$             'Selects free-running mode
PRINT #1, "$FR 0": INPUT D$             'Selects response
PRINT #1, "$SS 120": INPUT D$           'Selects scan speed
```

```
PRINT #1, "$GX"+ STR$(EX): INPUT D$
                                    'Sends monochromator to selected
                                    'wavelength
PRINT #1, "$MX 1": INPUT D$         'Selects emission monochromator
                                    'for scanning
PRINT #1, "$AL"+STR$(EM): INPUT D$
                                    'Sets emission low wavelength
PRINT #1, "$AH" + STR$(EMEND): INPUT D$
                                    'Sets emission high wavelength
PRINT #1, "$SC 1": INPUT D$         'Starts scan
    FOR I = 1 TO NUMBERPTS
    INPUT #1, VALUES(I)
    NEXT I
    PRINT #1, "$MA"                 'Returns to manual mode
```

The data are preceded by the header information, consisting of 10 numbers that indicate the status of the instrument (see Table 7.8). The data array contains the raw data that must be converted using the ordinate scale factor $(X(7))$ transmitted in the header information.

7.6.   AUTOMATIC BALANCES

Weighing, the most basic operation in analytical chemistry, can be automated using an articulated arm and a balance provided with communi-

Table 7.8. Header Information of Data Acquired from an LS-5

Digit	Meaning
$X(1)$	Instrument revision number
$X(2)$	Flags[a]
$X(3)$	Excitation wavelength
$X(4)$	Emission wavelength
$X(5)$	Scan speed, response
$X(6)$	Recorder format
$X(7)$	Ordinate scale factor
$X(8)$	Ordinate chart offset
$X(9)$	Number of data points
$X(10)$	Number of scans remaining

[a]The flags have the format WXYZ, where: W = 0 excitation, 1 emission, 2 both; X = 0 normal, 1 LC, 2 LP modes; Y = 0 fluorescence, 1 phosphorescence, 2 integrated phosphorescence; Z = 0 autorange, 1 fixed scale, 2 autoconcentration.

cation facilities. The bidirectional data output option of the Mettler AE balances is described next. These balances are currently being incorporated into the hardware and software of the ZYMARK laboratory robots. On the other hand, Mettler has also developed a balance specially designed to be operated by a robot. Thus, for instance, the protecting box may be removed to eliminate obstacles to the balance plate.

Mettler AE balances transfer weighing results to a data receiver (computer, terminal, etc.) via two data interfaces (20 mA current loop and RS232C). At the same time, however, they can also receive and carry out instructions. The RS232C data interface is set up as DCE (Data Communication Equipment). The standard 25-pole D-socket (female) has the following pin definition:

Pin 1   Protective ground
Pin 2   TxD, receive line
Pin 3   RxD, transmit line
Pin 4   RTS, request to send
Pin 5   CTS, clear to send
Pin 6   DSR, data set ready
Pin 7   GND, signal ground
....
Pin 20   DTR, data terminal ready

The communication protocol is 2400 bauds, even parity, 1 start and 1 stop bits, 7 bits per character, plus 1 parity bit. Each instruction must always be terminated by the character sequence CARRIAGE RETURN (CR, CHR$(13)) and LINE FEED (LF, CHR$(10)). In addition, all information transmitted from the balance to the data receiver is also terminated with the same character sequence, CR and LF. Every measuring result is transferred in a string, which can be divided into three blocks and which is also terminated with CR and LF (slashes added):

II/SP/DDDDDDDD/SP/u/CR/LF

- II is the identification block, which is composed of two characters. This block is used for exact identification of the transferred results.
- SP are spaces.
- DDD . . . is the data block, which contains the actual reading and which consists of nine characters. The reading is transferred into the D-block with sign in front, decimal point, and the particular applicable number of places.
- u is the unit block, which contains the symbol "g" for grams.

The instructions (which must alway be ended with CR LF) for control of AE balances are:

"S" (Send Value)

By using this instruction, the controlling instrument can request single measuring values via the data interface. After receipt of "S", the balance transmits the next stable weight result. The moment of transfer to the data receiver can be recognized by the brief blinking of the balance display. If the balance cannot make a meaningful weighing result available, it transmits "SI" (Send Invalid) at this point. This identification occurs, for example, when the weighing pan is at overload.

"SI" (Send Immediate Value)

For dynamic measuring, it is possible to make a request to a balance for weighing results that have not been released by the stability detector. A measuring result is transferred at the conclusion of the current display cycle by the instruction "SI". To differentiate dynamic from stable measuring results, the identification block contains the character sequence "SD" instead of "S".

"T" (Tare)

A tare procedure can be triggered by the instruction "T". If an attempt is made to trigger tare while the balance is in overload/underload range, the error message "EL" is sent.

"D text" (Display Text)

With this instruction, "text" messages of a maximum of seven characters may be displayed in the balance display. Missing text, that is, the "D " instruction (with a blank after the D), is interpreted as a blank text and leads to a balance display that is blanked out. Contrary to this,

	"D" (without blanks or text) returns to the normal display mode (display weighing results).
"C" (Clear)	This instruction has the same effect as switching the balance on/off.

In certain situations, the balance is not in a condition to carry out a received instruction. In such cases, the instruction received is rejected. To let the instruction transmitter know about this, the balance transmits an error message via the data interface. The message may be:

"ET" (Transmission Error)	Incorrect baud rate, parity, stop bits, and so on.
"ES" (Syntax Error)	Incorrect instruction set.
"EL" (Logistic Error)	Instruction sequence too fast, or more than seven characters sent to display.

Let us examine an example of how a Mettler AE balance may be handled by a computer:

```
OPEN "COM2:2400, E, 7, 1, LF" FOR RANDOM AS #1
. . . . . . . . . . . . . . . . . . . . . . . . . . . . . . . . . . . . . . . . . . . . . . . . . . . . . . . . . . . . . . . . . . . . . . . . . . . . .
PRINT #1, "T"              'To tare
PRINT #1, "S"              'Request a weighing operation
. . . . . . . . . . . . . . . . . . . . . . . . . . . . . . . . . . . . . . . . . . . . . . . . . . . . . . . . . . . . . . . . . . . . . . . . . . . . .
INPUT #1, WEIGHT$          'Read the weight
```

We developed the program BALANCE for automatic weighing and management of weight files. The main and auxiliary menus are shown in Table 7.9. On option, the weights are read, printed, and stored simultaneously to avoid accidental loss of the data, which can potentially ruin hours of work. For the same reason, the program contains a subroutine for error detection, to avoid situations that can lead to loss of data, such as an open diskette unit door, lack of space in the disk, or file name not found.

7.7. BURETTES AND DOSIFIERS

Automatic burettes and dosifiers are provided with interchangeable syringes with a piston driven by a micrometric screw. Modern burettes and

Table 7.9. Main and Auxiliary Menus of the Program BALANCE

Main Menu	Auxiliary Menu
1. Weight	1. Modify weight
2. Time	2. Erase weight
3. Date	3. Insert weight
4. Exit to DOS	4. Go to main menu
5. List file	5. List file
6. Modify file	6. Arrange file
7. Print file	7. Print file
8. Erase file	8. Exit to DOS
9. Merge file	

dosifiers have a devoted microprocessor that controls the movement of the piston, as well as a three-way valve that allows a shift from filling to delivery operation. Usually, these burettes can be optionally provided with standard interfaces for communication.

The Crison 738 autoburette can be controlled manually through the front panel or switched for remote control. The RS232C double interface allows connection of the burette as an intermediate element of a serial communication chain. The devoted microprocessor controls the valve and the syringe, the latter by means of a stepper motor. A complete displacement of the piston corresponds to 1000 steps of the motor. The communication protocol is the same given for the Crison 501 and 517 potentiometers. Control of the burette is performed through the internal instructions, which are written in a nonerasable memory of its microprocessor. These are:

0 to 9	Chain element number for request or control.
I	INPUT, sets the valve open to the left (reagent container, see Figure 7.6).
O	OUTPUT, sets the valve open to the right (delivery tube).
P	PICK UP, descendent piston direction to fill up the syringe. This command is generally used with the I instruction.
D	DISPENSE, ascendent piston direction to deliver the indicated volume. This command is generally associated to the O instruction.
XXXX	Digits (from 0 to 9) to indicate the step numbers of the motor for a burette addition:

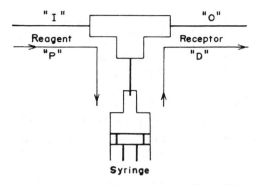

Figure 7.6. Meaning of the instructions I, O, P, and D of the Crison 738 burette.

$$\text{number of steps} \ = \ \frac{\text{volume to be added}}{\text{syringe volume}} \ \times \ 1000$$

	Example: 200 steps have to be performed to add 200 µL with a 1 mL syringe.
S	Autoburette speed selection, from 0 for the highest speed to 9 for the lowest.
F	To test if the burette is working.
Z	To test if the burette has reached an end of run position. A "Y" echo confirms this to the computer.
CR	CARRIAGE RETURN, executes a command transmitted by the computer.
C	Cancels an order transmitted by the computer when the instruction CR has not yet been sent.

The burette responds with an echo to the instructions or requests of the computer. The possible responses are:

X	A digit that corresponds to the identification number of the burette in the serial communication chain. This digit precedes all burette responses.
*	Indicates that the burette is still busy, performing some previous instruction.
?	Unacceptable command; there is a code the burette cannot understand.

Y   Affirmative response to question F or Z.
N   Negative response to question F or Z.

In order to control the burette as an element of a chain, the computer must fulfill the following protocol:

1. Open the communication channel.
2. Number the elements of the chain by sending an "N".
3. Control the element with the appropriate commands. For the Crison 738 burettes, these commands are:

"X"	Identification number, from 0 to 9.
"SX"	S stands for speed and X is a digit from 0 to 9 which establishes the syringe speed.
"I" or "O"	Sets the appropriate valve position.
"P" or "D"	Fixes the appropriate piston direction.
"X:X"	Number of motor steps which establishes the addition volume.
CR	Carriage return (CHR$(13)), to execute a command.

For example, when the burette is the first element of chain 1 (i.e., "0"), for the addition of 2 mL with a 5 mL syringe at the lower speed (i.e., "9"), the string sent by the computer must be

PRINT #1, "0S9OD400"

Since a semicolon is not given at the end of the string, the computer automatically includes a carriage return character and the instruction is immediately executed.

7.8.   SAMPLERS

7.8.1.   The Crison MicroSAMPLER 2040

The introduction of a sampler in an automated system is an important qualitative step because it can reduce the amount of time and attention that the user must devote to the system. Two samplers are described next—the Crison microSAMPLER 2040 and a homemade sampler based on the Fischertechnik general purpose interface (see Chapter 5).

The Crison sampler was designed to fulfill the requirements of most electrical methods. It is provided with 15 cell locations in a circular carousel. The cells are large (about 50 mL), with wide mouths, and the cell located at the measuring position can be raised for the introduction of the electrodes, thermometer, stirrer, and so on.

The sampler has an RS232C interface with the same specifications as the Crison 501 and 517 potentiometers and the 738 burettes; it can therefore be chained with them. However, since only one interface is available, the sampler can only be located at the end of the chain.

The sampler instruction set is:

P	Pick up, the cell in the first position is raised.
D	Dispense, the cell is descended.
N	Numbering, the sampler echoes its position number in the chain. This number must precede any other command.
F	Tests if the sampler is working. The echoed response may be "Y" or "N".
Z	Tests if the sampler is in position 0 (cell descended) or 1 (cell raised).
C	Clear, cancels an order if it has not yet been executed by a CR.
CR	Carriage return, executes a command.
A	After, turns the carousel one position clockwise.
B	Before, turns the carousel one position counter-clockwise.
GXX	Get, cell XX is located in the measuring position.
L	Activates a washing cycle of the electrodes.
TXXXXYYYY	Format to program a washing cycle. The first two Xs indicate the position of the first cell containing washing solution. The third and fourth Xs indicate the number of seconds this cell must be raised. The Ys have the same meaning with respect to a second cell with washing solution.
E	To control the start and stop of the stirrer.
S	Status, the sampler responds as follows (spaces and upper digits added):

$$\begin{array}{ccccccccccccc} 1 & 2 & 3 & 4 & 5 & 6 & 7 & 8 & 9 & 10 & 11 & 12 & 13 \\ C & X & X & X & X & X & X & X & X & X & X & X & X \end{array}$$

where C is the cell status; it may be P (raised) or D (descended); digits 2 and 3 indicate the number of the cell located at the measuring position; digits 4 and 5 indicate the total number of cells to work with; digits 6 and 7 show the position of the first washing cell; digits 8 and 9 show the duration of the first washing operation. The meaning of digits 10–13 is the same, with respect to the second washing cell.

Thus, for instance, a typical response may be (spaces added)

P 01 15 14 10 15 05

which means that the first cell is raised (P), all the cells contain solutions to work with (15), the washing solutions are located in cells 14 and 15, and the washing operations will last 10 and 5 s, respectively.

The sampler responds with an echo to all the received instructions or commands. Possible echos are:

*   Busy, a previous instruction is still being executed.
?   Unacceptable command. The sampler cannot understand a received code.
Y   Affirmative response to F or Z.
N   Negative response to F or Z.

7.8.2. A Homemade Sampler Based on a Fischertechnik Kit

The Fischertechnik general purpose interface (see Section 5.6.3) provides a very easy way of constructing and programming an automatic sampler. The sampler described here was constructed and used in the authors' laboratory with FIA and atomic absorption setups with excellent results.

A scheme of the sampler is shown in Figure 7.7. The Fischertechnik interface is used to control two dc motors and three interruptors. The carousel is also circular, but the cells are not raised. Instead, the tube for aspiration of the sample is introduced into them by means of one of the motors, which moves along a toothed vertical bar. The end of run positions are detected with two interruptors located in regulable fixed positions on the bar. The carousel is supported by a toothed disk. The disk is turned by a motor by means of a toothed wheel. An interruptor is used to control the turn position of the carousel. The interruptor counts the number of teeth opening and closing it.

To aspirate a sample, the tube tip is lowered until the lower interruptor is closed. The time of aspiration is software controlled. When the specified time has elapsed, the tip is raised until the upper interruptor is closed. Then,

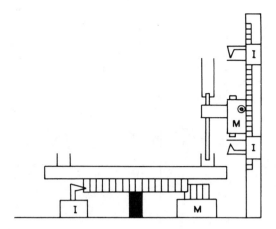

Figure 7.7. Scheme of a homemade sampler.

the motor that moves the carousel is activated until a specified number of teeth have been counted, and the cycle begins again.

Programming is facilitated by the INTERFAC.COM subroutine, which is supplied with the Fischertechnik interface. The BASIC examples given by the manufacturer, with a few simple modifications, may be used to control this or any other machine prototype.

CHAPTER

8

AUTOMATION OF ANALYTICAL METHODS

8.1. INTRODUCTION

Automation of analytical methods offers several important advantages, which may be summarized as follows:

1. Since computers are tireless and have no opinions or preferences, the criteria given to prepare the sample and to take the measurements are objectively followed during the whole experiment.
2. Since the experiments can be performed without supervision overnight, the researcher is only needed to initiate the operation.
3. It frees the researcher from the tedious tasks, allowing him or her to dedicate more attention to the experimental design and interpretation of data.
4. More information can be obtained, since more points can be taken, more experiments can be carried out, and the conditions can be changed more extensively.

The automation of several analytical methods and procedures is described in this chapter. All the corresponding automated systems were developed by the authors using the devices and pieces of equipment described in previous chapters. This chapter is therefore a collection of examples showing how automation sciences may be applied to the chemical laboratory to perform both routine and research tasks. All programs cited in this chapter are available at SCIWARE (registered trademark of Association of Environmental Sciences and Techniques, Department of Chemistry, The University of the Balearic Islands, E-07071 Palma de Mallorca, Spain).

Due to the interests of the authors, stress is placed on analytical instrumentation and procedures. However, many other nonanalytical applications, such as the automatic regulation of experimental conditions (e.g., pH, pressure, temperature, concentrations) in synthesis or biochemical processes, may easily be implemented making use of the examples given here.

8.2. AUTOMATION OF POTENTIOMETRIC TITRIMETRY

8.2.1. Introduction

Potentiometric titrimetry is widely used as both a routine analytical method and as one of the most accurate and precise methods for the evaluation of stability constants. The latter usually requires waiting for the electrode to reach equilibrium after each titrant addition before taking a measurement. Times as long as several hours are frequently required near the equivalence points (nonbuffered regions) and sometimes even over the whole titration (e.g., acid–base titrations with a glass electrode in DMSO). This makes the experimental procedures tedious and time consuming and the measurements are not always objectively taken. Therefore, potentiometric titrimetry can benefit greatly from automation.

Several automatic potentiometric titrators have been marketed during the last decade. They have proved to be very useful in routine tasks but not in research, due to their closed design. Most of them cannot be adapted by the user to the particular requirements of many research tasks.

8.2.2. Automation of Potentiometric Titrations Using a Commodore VIC-20 Microcomputer

The Commodore VIC-20 microcomputer, with only 3.5 kbytes of RAM memory, is enough to implement an automatic system for potentiometric titrimetry. The program VALPOT was designed by the authors to be run with the minimum configuration of the VIC-20 and works equally well with the more powerful Commodore 64 computer. Other elements of the system were a Commodore MPS 801 printer, a Commodore C2N cassette tape driver, a B/W TV monitor, a Crison Digilab 517 potentiometer, and a Crison 738 autoburette. The two latter instruments were provided with RS232C interfaces. The block diagram is shown in Figure 8.1.

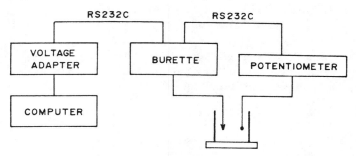

Figure 8.1. Block diagram of an automatic system for potentiometric titrimetry based on a VIC-20 microcomputer.

Most microcomputers have a UART, which is able to communicate with the exterior. A UART can transform serial transmitted data into words of parallel information and vice versa. It can also program the speed of the input and output data in order to enable interface compatibility.

When production of the VIC-20 microcomputers began, the convenient UART was not available, and Commodore decided to emulate an interface by software using the VIA 6522 (Versatile Interface Adapter), compatible with its 8-bit microprocessor 6522. The user of VIC-20 can directly access the VIA lines through the user port. However, because the VIA has TTL technology, its direct connection with the standard RS232C interfaces is hindered. This difficulty may be overcome with a simple voltage adapter (see Figure 6.2, the MC1488 and MC1489 chips give an adequate line voltage).

Due to its memory limitations, the VIC-20 was used only to control the experiment and to acquire the data. The data treatment was always performed with another computer (an HP 85/86) using the data files stored on a cassette tape.

8.2.3. Automation of Potentiometric Titrations Using an IBM PC

Apparatus. An IBM PC microcomputer with 256 kbytes of RAM and graphic screen card, a printer, a Crison 738 automatic burette, and a Crison Digilab 517 potentiometer provided with Orion combined glass and silver 94-16A electrodes (with a Crison electrode adaptor) were used. The Crison instruments were serially connected to the PC through RS232C interfaces. The modular system is shown in Figure 8.2. Since the external ports of the portable IBM PC microcomputer work at the RS232C standard potential, direct connection to the instruments is possible.

Software. The logical support of the system is provided by the program VALGRAN, which was developed to meet the following criteria:

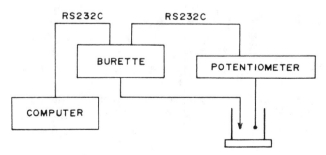

Figure 8.2. Block diagram of an automatic system for potentiometric titrimetry based on an IBM PC.

1. No specific training is required to run the program, which is driven by very simple menus.

2. The titration rate automatically meets the requirements of the experiment.

3. Printing of the data and graphic representation of the titration curve on the monitor screen are carried out simultaneously with the experimental task. In this way, any irregularity can be detected immediately and better quality control can be achieved.

The main menu contains the following options:

1. *Titration for Chemical Equilibrium Studies.* In this case, the electrode must reach equilibrium between successive additions of titrant. For this purpose, a fixed waiting time is required between consecutive potential measurements. A measurement is accepted for storage only when the difference from the previous one is smaller than a given value. When a measurement has been accepted, a new addition of titrant is performed and another sequence of measurements is carried out.

This strategy permits determination of the stability constants of many systems with a high degree of accuracy and precision, even when a very long time is necessary for the solution–electrode equilibrium to be reached.

When a titration is finished, the data are stored in a diskette file. By means of a word processor, these files can easily be made compatible with many parameter optimization programs such as PHCONST or the well-known MINIPOT, MINIQUAD, or SUPERQUAD.

2. *Titration for the Determination of Equivalence Points.* A different criterion must be used to control the titration rate in routine work, since stabilization of the electrode potential in the buffered regions is not necessary. In this case, a compromise between accuracy and time may be established. To achieve a high titration rate, in the program VALGRAN only a fixed short waiting time is allowed between the end of an addition and the measurement of the potential, but the volume added is automatically modified inversely to the slope of the titration curve. Far away from the equivalence point, the slope is small, and large volumes are added. Near the equivalence point, increasingly higher slopes are found, and thus smaller volumes of titrant are used. Finally, after the equivalence point, the volume added increases again. In this way, the system works as a skilled analyst would, and accurate determinations can be performed in 2–3 min.

In addition, if some care is put into the software design, a considerable amount of time may also be saved. Thus, at the beginning of the titration, some delay time is needed until the burette is full up, but there is a considerable waste of time between consecutive additions if a fixed delay time is employed. Instead, the current state of the burette may be continuously checked during the filling up operation. When the plunger stops moving, the program understands that the burette is full up and the titration begins immediately.

Furthermore, because of the difference in execution speed between interpreted and compiled programs, the use of FOR-NEXT based idle delay loops (e.g., FOR I = 1 TO 10000 : NEXT) is not advisable. In the program VALGRAN all the delay loops are clock controlled (using the TIMER function).

The possibility of the equivalence volume being larger than the syringe burette volume has also been considered. When the syringe is empty, it is automatically refilled. If the end-point has not been found after a given number of fillings, a message is issued indicating that dilution of the sample is advisable.

3. *Other Options.* Other options include manual introduction of the data (through the keyboard), retrieval of stored data (from a diskette), and return to DOS.

The submenu for data treatment contains the following options:

(a) Representation of the titration curve on the monitor screen.

(b) Calculation of the equivalence volume and concentration of the sample by means of the second derivative method. Also, as an option, the first and second derivative curves can be plotted on the screen.

(c) Calculation of Gran's functions, before and after the equivalence point, by linear regression. The two regression straight lines and the corresponding experimental points are plotted on the monitor screen. In this way, the regions containing the best experimental points can easily be selected and the regression analysis repeated if necessary. Independent values of the equivalence volume, as well as the two standard electrode potentials, before and after the equivalence point, can be obtained from the crossing points of the two straight lines with the X-axis. For acid–base titrations, the values of the standard potentials allow calculation of the ionic product of water under the conditions of the experiment.

The program VALGRAN was tested with several acid–base and chloride–silver titrations with satisfactory results.

8.2.4. Automatic Determination of Boric Acid

The determination of boron is frequently performed in many samples of industrial interest. We developed an automatic system to perform part of the involved analytical procedure. The system is being used on a regular basis for determination of boron in ceramic materials.

The samples are first melted down with sodium carbonate and the residue is acidified with a small HCl excess. The automatic system titrates the HCl excess with NaOH standard solution, adds mannitol to enhance the acidity of boric acid, and, finally, resumes the titration with the same titrant, until a second equivalence point is reached.

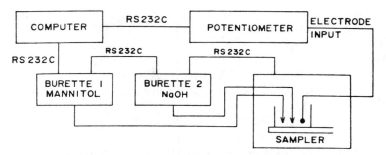

Figure 8.3. Block diagram of the automatic system for the determination of boric acid.

Apparatus. A diagram of the system is shown in Figure 8.3. An Acer 710 compatible computer, provided with 640 kbytes of RAM memory, hard disk and floppy disk drivers, a Crison 2002 potentiometer/pH meter, two Crison 738 autoburettes, and a Crison 2040 autosampler were used. All these instruments were originally provided with the corresponding RS232C interface, and a second RS232C interface was installed in the computer to control the potentiometer. The latter needs to be independently controlled, since its communication rate is 1200 baud, whereas the burettes and the sampler communicate at 2400 baud. Samples were located in cells 1–13 of the autosampler, the last two cells being reserved for the electrode washing solutions.

Software. Logic support of the system was provided by the program BORIC. The main menu of the program contains the following options:

1. Perform experimental task.
2. Retrieve and treat data files.
3. End the process.

When option 1 is selected, a submenu with the default working conditions appears: for example, titrant concentration, volume of mannitol, burette syringe volume, titrant addition volumes, maximum titrant volume, and number of samples to be titrated. Either, the default conditions or a previously stored file containing other working conditions may be used. New files with different working conditions may also be created and stored.

An experimental curve is shown in Figure 8.4. To perform a titration, the electrodes are washed and then located in one of the cells reserved for the samples. The sequence of sample cells is automatically selected. The excess of HCl is first neutralized. The equivalence point is established as the maximum of the first derivative of the titration curve. The corresponding titrant volume is set as the new starting point, and a given volume of mannitol is

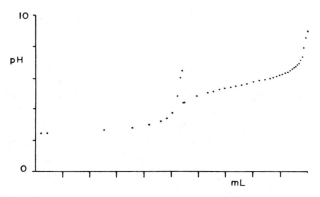

Figure 8.4. Experimental curve showing the two equivalence points, before and after the addition of mannitol. Under the plot the following messages (not shown) appear: equivalence volume = 2.19 mL; press C to continue and M to return to the menu.

added. If pH does not decrease, the titration ends. If it does, the titration is continued until a new maximum of the first derivative is reached. The new volume added is used to calculate the contents in B_2O_3 in the original sample.

Simultaneously with the experimental work, the titration curves appear on the monitor screen, which gives the researcher an overview. At the same time, the data are stored in a diskette file, and the final values of the titrations are printed, which avoids the irreversible loss of information that could result if a power failure occurs. Also, the final analytical report is created, stored, and optionally printed.

8.2.5. An Automatic System for Ion-Selective Potentiometry

An automatic system for ion-selective direct potentiometry and potentiometric titrations is described next. The system incorporates a sampler that automatically provides calibration curves and analyzes a high number of samples without supervision. Preparation of the sample with buffers, surfactants, and so on may also be performed.

Apparatus. The block diagram of the system is the same as that used for the determination of boric acid (see Figure 8.3) with the difference that the burettes may contain other solutions. Thus, in the application examples that were used to check the system, the burette containing mannitol was filled up with a TISAB (total ionic strength adjustment buffer). Ion-selective electrodes for chloride and lead ions from Orion and for potassium, sodium, and ammonium ions from Ingold were used.

Software. The main menu of the program ISE, designed to support the system, contains the following options:

1. *Direct Potentiometry.* This is split into two options: ordinary calibration curve and application of the standard addition method. The submenus contain options to select the cells that will contain the standards (for direct potentiometry), washing solutions, and samples, as well as to establish the burettes that contain the standard (for titrimetry) and the buffer, or other solution for sample conditioning. The experimental conditions that define each application procedure can be stored in a file that can be called when needed.

When all the standards have been measured, the calibration curve is established and the samples begin to be measured, the concentrations calculated, stored, and optionally printed. The printout table contains the sample number (cell position in the sampler tray), its potential, concentration, and electrode stabilization time. All results are stored in a file that may be retrieved to repeat calculations and/or to plot the curves if needed. The standard addition method is similarly applied.

2. *Potentiometric Titrations.* Titrations are performed in the way described in Section 8.2.3 (program VALGRAN). The equivalence volume is calculated from the point were the second derivative is zero.

3. *Gran's Method.* Constant amounts of titrant are added after the equivalence point. Gran's function is used to calculate the sample concentration.

8.2.6. Automatic Potentiometric Titrimetry with Electrogenerated Reagents

In this system the titrant is electrochemically generated. This is an example of how a computer may be easily used to implement a less common analytical technique at a very low cost.

Apparatus. The block diagram of the system is shown in Figure 8.5. An IBM PS/2 model 30 computer was provided with two cards: a PC ADDA-14 FPC-011 board (from FLYTECH Technology Co., Ltd.) and a FC-046 P Industrial I/O board. Other elements of the system are a sampler, a homemade galvanostat, and an electrolytic cell provided with two pairs of electrodes. One of the pairs is constituted by the reagent source electrode (W) and an auxiliary counterelectrode (A) for reagent electrogeneration (e.g., a silver wire and a platinum mesh electrode for halide titrations). The other electrode pair is constituted by an indicator (I) and a reference electrode (R). This pair allows monitoring of the titration along the galvanostatic (constant intensity) electrogeneration.

We used an Orion silver-ion-selective electrode for halide titrations and a Hanna HI 1911B combined glass electrode for acid–base titrations. The latter electrode is provided with a high input impedance amplifier with a mercury battery to supply the amplifier, all of which is incorporated into the

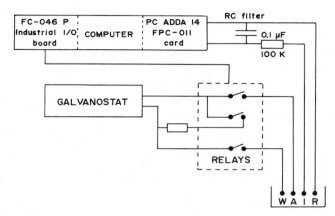

Figure 8.5. Block diagram of an automatic system for potentiometric titrimetry with electrogenerated titrant. W=working electrode, A=auxiliary electrode, I=indicator electrode, R=reference electrode.

electrode rod, which affords high stability and low output impedance. This allows direct connection of the electrode to any inexpensive digital voltmeter or A/D converter, avoiding the use of a very high input impedance potentiometer. Finally, an RC filter at the input of the A/D converter is used for better stability of the readings.

Software. The program COULOM works as follows. The value of the constant current for reagent generation is previously established. The electrodes are first introduced in the washing solutions and then in the selected sample. When the computer closes the relays that connect the galvanostat with the reagent source and counterelectrodes, the time counting and the data acquisition process begin. The time–potential values are plotted on the monitor screen and stored in a file. The process ends when the prefixed time has elapsed. Then, the equivalence point is obtained as the maximum of the first derivative, and the experimental conditions and initial data are used to calculate the sample concentration.

The program has been applied to the determination of chloride in water, wine, beer and dairy products, as well as to the determination of the acidity and basicity of environmental and food samples.

8.3. AUTOMATION OF POTENTIOMETRIC STRIPPING ANALYSIS

8.3.1. Introduction

Potentiometric stripping analysis (PSA) is a modern analytical technique introduced by D. Jagner and co-workers for trace and ultratrace analysis. In PSA, the working electrode is first driven to a sufficiently low potential to reduce the metals of interest. This is the pre-electrolysis step, in which the metals are concentrated on the electrode surface during a prefixed time. In a second step, the working electrode is disconnected from the potential source and remains connected only to a high input impedance voltmeter.

The metals are reoxidized by an oxidant present in the unstirred solution and the potential–time curve is monitored. The potential of the electrode increases quickly until the Nernst potential for the oxidation of one of the deposited metals is reached. The redox buffer of the metal keeps the potential almost constant until depletion of all the reduced form of the metal has been completed. Then, the potential rises again until the Nernst potential of another metal is reached. Thus, each metal produces a plateau at a specific potential value. The length of the plateau is proportional to the amount of metal.

One of the main advantages of PSA is that ordinary and very simple instrumentation is required: a general purpose potentiostat and a paper chart recorder. Practical application of PSA is tedious when very low metal levels have to be determined, since the pre-electrolysis step requires a very long time. Automatic techniques are therefore very convenient for PSA. We developed two semiautomatic systems for PSA, the first one based on a Commodore VIC-20 microcomputer and the second one based on an IBM PC. The latter has the ability to perform the stripping step by electrochemical means.

8.3.2. Automatic System for PSA Based on a VIC-20 Microcomputer

Apparatus. The block diagram of the system is shown in Figure 8.6. The Commodore VIC-20 microcomputer was provided with 16 kbytes of memory expansion, a SUPEREXPANDER cartridge for graphic applications, and a diskette unit. The RS232C serial interface was emulated through the user port, which was used to control the burette and the potentiometer and

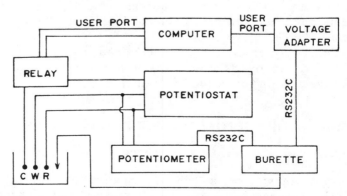

Figure 8.6. Automatic system for potentiometric stripping analysis based on a VIC-20 microcomputer.

simultaneously trigger the control relays by means of optocouplers (see Figure 2.24). Since the emulated RS232C serial interface of the VIC-20 computer works at TTL voltage levels, a voltage adapter was needed to interface it to the RS232C of the Crison potentiometer.

A general purpose homemade potentiostat, an electrolytic cell provided with a reference standard calomel electrode (R), a glassy graphite working electrode (W), and a Pt auxiliary counterelectrode (C), a Crison 517 potentiometer, a Crison 738 automatic burette, and a homemade board supporting several control relays were also used. The Crison instruments were provided with RS232C interfaces.

The computer starts and stops the pre-electrolysis according to a predefined time. The potentiometer transmits the acquired data to the computer. The potential variations are monitored during both the pre-electrolysis and stripping steps. The burette permits automatic application of the addition standard method.

Software. The program PSA was split into two parts—PSA1 and PSA2—due to the limited memory of the computer. The program PSA1 carries out the experimental functions: data acquisition and storage, connection and disconnection of the electrodes, addition of the standards, and so on. It performs a predetermined number of additions of the standard solutions.

The program PSA2 is used for data treatment. It identifies the components of the sample, plots the calibration curves, and calculates the length of the plateaus for the series of standard additions and, hence, the sample concentrations.

Control of the Burette. An example of how an addition of titrant is performed is given next:

```
. . . . . . . . . . . . . . . . . . . . . . . . . . . . . . . . . . . . . . . . . . . . . . . . . . .
XX10 OPEN 2,2,0,CHR$(170)+CHR$(112)
                         'Establishes the communication channel
XX20 PRINT #2, "N"       'The elements of the chain are numbered
. . . . . . . . . . . . . . . . . . . . . . . . . . . . . . . . . . . . . . . . . . . . . . . . . . .
XX90 V$= STR$(VA/VJ*1000)   'VA = addition volume, VJ = syringe
                            'volume (equivalent to 1000 motor steps)
X100 PRINT#2, "0S1OD" + V$ + CHR$(13)
. . . . . . . . . . . . . . . . . . . . . . . . . . . . . . . . . . . . . . . . . . . . . . . . . . .
```

Control of the Relays. The user's port of the computer is used to close and open the connection between the electrodes and the potentiostat. The I/O lines of the port are controlled by means of two memory addresses: the port at address 37136 and the data address register at 37138. At the port, an input datum is read (acquired) with PEEK and an output datumm is written (sent) with POKE. The data address register is used to establish the status of each port pin for input or for output. Thus, for instance, to send a signal through pin E (PB2), a possible BASIC sequence would be the following:

XX10 POKE37138, 255   'Establishes all port pins for output
XX20 POKE37136, 4   'Activates pin PB2 sending a 4 to the
                           'bus memory address

In binary a 4 contains a bit in logic 1 (0000 0100). The corresponding digital signal or digital pulse is enough to activate the relay.

Synchronization of Reading Cycles. The potentiometer transmits the data to the computer through the serial interface. A difficulty is to control the delay time between the computer called and the response of the potentiometer, which is not constant. A convenient method for synchronization is to implement a loop in which the characters are continuously read, until a plus ($+$) or a minus ($-$) sign arrives. The sign character is thus used as an indication that a new datum begins.

Example

. . . .
XX50 PRINT#2, "1"   'A datum is requested from the potentiometer
                           '(element "1")
XX60 GET#2, C$: IF C$=" " THEN XX50
                           'A datum is requested iteratively until a
                           'response is obtained
XX70 W= 0   'Initiates a safety count to avoid a closed
                           'endless loop
XX80 GET#2, C$
XX90 IF C$= "+" OR C$= "−" THEN B$= C$: GOTO X120
X100 W= W+1 : IF W= 30 THEN XX50
X110 GOTO XX80   'If a + or − sign has not been received,
                           'another character is read
X120 FOR V= 1 TO 4
X130 GET#2, C$
X140 B$= B$ + C$
X150 NEXT V
. . . .

This method is rather slow but yields reliable readings.

Data Treatment. The data treatment presents a number of problems, which we solved as follows. The first problem is to automatically detect the plateaus of the potential–time curves, which may be achieved by means of the second derivative curves. The potential–time curves show a sudden slope increment before and after each plateau, which implies a sign change of the second derivative. The program detects these sign changes to establish the beginning and the end of each plateau.

However, the presence of noise associated with experimental measurements also produces sign changes of the second derivative, which does *not* correspond to the transition plateau from one element to another. In the program PSA, two methods are used to discriminate the noise. First, when the program detects a sign change with the second derivative method, it requires a high value of the first derivative before considering the sign change as the beginning or end of a plateau. Second, the potential of the possible plateau is compared with the potentials of the elements to be determined, which are stored in a file. The plateau is not considered if its potential does not agree with any in the file. In this way, good results are obtained, even in the presence of a high level of noise.

The calculation of the length of the first plateau presents an additional problem. When the pre-electrolysis is disconnected, the potential suddenly increases, until the level of the first plateau is reached. If data begin to be acquired immediately after disconnection of the pre-electrolysis step, the rapid increase of the potential hinders the acquisition of enough data to calculate the length of the plateau. However, satisfactory results are obtained by beginning the data acquisition process a few seconds before disconnection of the pre-electrolysis.

Once the plateaus are localized, linear regression is applied to the points of the plateau and to the zones before and after the plateau. The points to be considered for linear extrapolations must be chosen manually by the user. The crossing points of the resulting straight lines are used to compute the length of the plateaus. Finally, the computer automatically calculates the concentration of the samples using the plateau lengths obtained by the standard addition method.

Preparation and Regeneration of the Electrode Surface. It is very important to maintain a reproducible electrode surface, at least during application of a series of standard additions. The glassy carbon electrode must be polished several times daily for 1–2 min and carefully washed with acetone to remove any trace of fat and fingerprints. Before each series of measurements (e.g., each application of the standard addition method), the electrode is covered with a mercury film, which has to be thick enough to dissolve all the mercury-soluble metals that are reduced during the plating period. This is achieved by holding the electrode at −0.5 V versus SCE for 10–30 s in a nondeaerated solution containing 0.1 M hydrochloric acid and 25 mg/L Hg(II).

A stripping curve is recorded to establish the base line. The plating potential is then changed to −0.60 V and the plating/stripping cycle repeated. This procedure is repeated at −0.70, −0.80, and −0.90 V. Finally, the 2-min plating period is initiated at −0.90 V, and after stripping the working electrode is rinsed with distilled water.

Analytical Procedure and Results. The standard addition method was applied to samples containing 0.05 M hydrochloric acid, 0.5 M sodium chloride, and 1 mg/L Hg(II). The samples were previously purged for 15 min with nitrogen, and a nitrogen stream was maintained over the sample surface during the electrochemical process. The sample was treated at −0.95 V for 5 min during the pre-electrolysis period. Different volumes of a standard containing Pb(II) and Cd(II) were added with the automatic burette and, each time, the plating/stripping cycle was repeated.

The PSA curve of the blank is shown in Figure 8.7a. The background noise was identified by the program as such, and it pointed out that there was no element to be determined. The curve corresponding to the addition

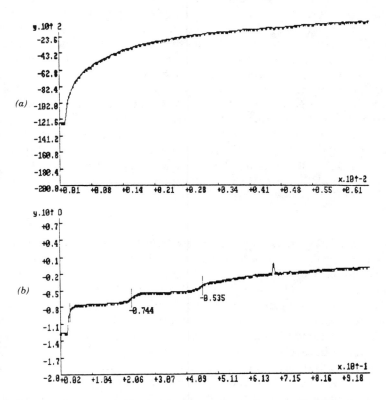

Figure 8.7. (a) PSA curve of a blank with 0.05 M HCl, 0.5 M NaCl, and 1 μg/mL Hg(II). (b) The same plus 250 ng/mL of Cd(II) and Pb(II).

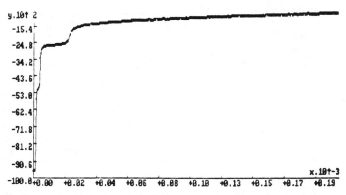

Figure 8.8. Determination of 1 ng/mL of Pb(II) with the automated PSA system. Experimental conditions are the same as in Figure 8.7. The small plateau at −0.51 mV is a Cd impurity.

of 250 ng/mL Cd(II) and Pb(II) is shown in Figure 8.7b. The program identified these two elements, calculated their potentials, and detected the beginning and end of each plateau.

Finally, Figure 8.8 shows the capability of the system for determining low concentration levels. The figure corresponds to the PSA curve of 1 ng/mL Pb(II) with a pre-electrolysis time of 2 h 30 min. The plateau was correctly detected and its length calculated. On the other hand, the presence of cadmium as an impurity was not detected by the program. However, due to the possibility of manual treatment, its quantification could be carried out.

8.3.3. Automatic System for PSA with Galvanostatic Oxidation Option Based on an IBM PC

Apparatus. A simpler and more powerful PSA system was built up using an IBM PC. Communications were implemented by means of a MetraByte DASH-8 card, which permits a higher data acquisition rate, and therefore makes it possible to study processes with very short plateau lengths, including the use of oxygen as the oxidant and short pre-electrolysis times. A block diagram of the system is shown in Figure 8.9. In comparison with the PSA system given above, the voltage adapter is not needed, and a galvanostat has been added to implement the electrochemical stripping option.

Software. Logic support of the system was provided by the program COMPSA (compiled program for PSA), which also includes several subroutines for data treatment. Compilation allows a higher reading rate and the use of the subroutine library of the DASH-8 board (DASH8.OBJ), which is linked with the user's program when it is compiled.

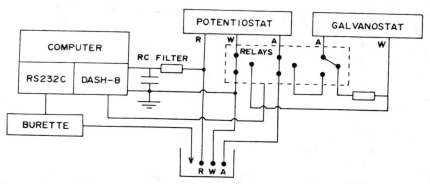

Figure 8.9. Automatic system for potentiometric stripping analysis with galvanostatic oxidation based on an IBM PC. R=reference electrode, W=working electrode, A=auxiliary electrode.

The program COMPSA, written in QuickBASIC, contains a number of options with menu presentation to define the working conditions in a similar way to the programs presented above. The data treatment subroutines are similar to those developed for the program PSA2 (see above).

Control of the Relays. Pre-electrolysis time is controlled by the computer through the DASH-8 interface by means of a series of relays that connect and disconnect the electrodes. Another relay is used for the independent connection of the galvanostat. To activate the relays (through optocouplers), the digital outputs of the DASH-8 card are used. This is facilitated by using the machine language subroutine MODE 14 of the card software:

```
MODE%= 14
DIO%= outputnumber
CALL DASH8 (MODE%, DIO%, ERRORS%)
```

where outputnumber is an integer within the range 0–15 which represents the 4 bit set to be transmitted through the digital port. Thus, for instance, to drive outputs OP1 and OP3 to high and outputs OP2 and OP4 to low, we should write DIO%= 5, which corresponds to the binary number 1010. The TTL level of +5 V that appears at the corresponding first and third digital outputs of the board is enough to trigger the relays.

Rapid Reading Routine. The program includes a subroutine for noise reduction. This is achieved by taking about 200–300 data at a high rate. The mean value is calculated and the result is used as a single point of the potential–time curve. This is a useful application of the MODE 5 subroutine of the DASH-8 board. A given number of evenly timed A/D conversions (readings), entirely under the subroutine control, are performed with a single instruction. Each conversion is performed when an upward edge (interrupt input) is detected at pin 24 of the board. These interrupt inputs are provided by an on-board timer. Thus, to implement the rapid reading routine, the output of one of the timers of the board (e.g., the CTR.2 OUT, at pin 6 of the board) is externally connected to the interrupt input (pin 24), and the timer is programmed to provide the adequate pulse train. For example, to generate 10 readings every 5 counts of the timer, the following BASIC sequence may be used:

```
MODE% = 10               'Initialization of the timer #2 in the operation
                         'mode 2 (pulse generation)
DIO%(0) = 2
DIO%(1) = 2
CALL DASH8 (MODE%, DIO%(0), ERRORS%)
MODE% = 11               'Sends to the timer #2 the operation parameter 5
                         '(a pulse every 5 counts)
DIO%(0) = 2
DIO%(1) = 5
CALL DASH8 (MODE%, DIO%(0), ERRORS%)
MODE% = 5
DIO%(0) = VARPTR(DATA%(0))
                         '10 readings are performed and stored in the
                         'vector DATA%(X)
DIO%(1) = 10
CALL DASH8 (MODE%, DIO%(0), ERRORS%)
```

8.4.   AN AUTOMATIC SYSTEM FOR CONDUCTIMETRIC TITRIMETRY

Apparatus. The block diagram of the system is shown in Figure 8.10. The Commodore VIC-20 microcomputer was provided with a 16 kbyte RAM expansion module (VIC-1111), a printer, a cassette tape for data storage,

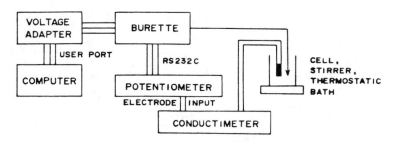

Figure 8.10. Block diagram of an automatic system for conductimetric titrations.

and a B/W TV monitor. A Crison Digilab 517 potentiometer, Crison 738 burette, and Crison 522 conductimeter with a Philips conductivity cell were also used. Only the potentiometer and the burette were provided with RS232C interfaces. The data taken from the recorder output of the conductimeter were digitalized by the potentiometer and input to the VIC-20 through an RS232C/TTL voltage adapter.

Software. The program VALCON was designed following the usual criteria of easy operation and multiple options to treat, store, and retrieve the data. It contains a subroutine for the automatic determination of equivalence points. For this purpose, regions of closely aligned points are sought, linear regression is applied, and the crossing points of the straight lines are calculated. The typical curved regions at the beginning of the titration curves are automatically rejected. On option, the limits of the intervals for linear regression can also be manually selected, with the help of the titration curve that is plotted on the monitor screen. This automatic system proved very useful for determination of mixtures of acetic and chlorhydric acids, SO_2 in polluted air, and sulfates in waters.

8.5.   AUTOMATION OF SPECTROPHOTOMETRIC METHODS

8.5.1.   Automation of Photometric Titrations

Apparatus. The block diagram of a system that performs automatic photometric titrations is shown in Figure 8.11. A Beckman ACTA III spectrophotometer provided with a 50 μL flow-through cell, a Crison Digilab 517 potentiometer, a Gilson Minipuls 2 peristaltic pump, a Commodore VIC-20 microcomputer, and a TTL/RS232C voltage adapter were used.

Software. The main menu of the program TITSPEC presents the following three options:

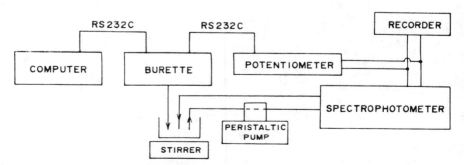

Figure 8.11. Block diagram of an automatic system for photometric titrations.

1. *Initialization*. The necessary data for sample identification and the working conditions are required. When all these have been supplied, the program opens the communication channel, initializes the printer, and fills up the burette.

2. *Experimental Task*. The titration is performed and the data are acquired. This is done through the potentiometer, which performs the functions of A/D converter and RS232C interface. A reading is iteratively requested until a correct format is achieved (the datum has been completely read). For a given addition of titrant, the data are iteratively acquired at a given rate until a satisfactory stability is reached. The criterion for stability is the coincidence of consecutive readings within the previously established limits. An accepted datum is stored and the titration continues until all the desired titrant volume has been added. The burette syringe is automatically filled up as many times as needed.

The simultaneous recording of the analog signal can easily be implemented with a paper chart recorder connected in parallel to the spectrophotometer recorder output. This provides a hardcopy of the data, which is useful in case of a power failure before storage of the acquired data.

3. *Exit*. Closes the communication channel and returns to DOS.

Results and Data Treatment. The system was used for the study of the Cu(II)–ethylenediamine complexes. The data files obtained with TITSPEC were directly transmitted from the VIC-20 to an IBM PC through a TTL/RS232C voltage adapter and an RS232C interface. Also, direct communication between the IBM PC and a VAX 750 computer from Digital was implemented with the program SMART TERMINAL, option KERMIT, of the VAX computer. The latter was used to evaluate the stability constants of the complexes, with an adaptation of the program MINISPEF.

8.5.2. An Automatic System for Spectrophotometric–Potentiometric Titrations

Protonation constants are frequently evaluated from absorbance–pH data. An automatic system for the control and data acquisition of titrations with simultaneous photometric and potentiometric monitoring is described next.

Apparatus. The block diagram is shown in Figure 8.12. The experimental setup is very similar to that of Figure 8.11, but a Crison Digilab 517 potentiometer and a combined standard glass electrode were added.

Software. The program POTSPEC contains many of the features of VALPOT and TITSPEC. A difference is that here the stability of the measurements must be reached simultaneously in both absorbance and

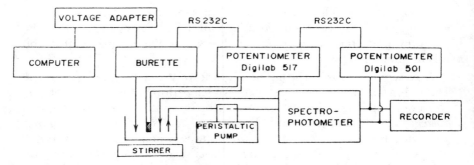

Figure 8.12. Block diagram of an automatic system for spectrophotometric–potentiometric titrations.

potential readings. This system minimizes many of the sources of error usually found in manually performed experiments (position of the spectrophotometric cell, volumes of added reagents, stabilization times, etc.). This leads to highly reproducible, high-quality results, which permit the data to be fit successfully to complex models.

8.5.3. Data Acquisition System for Spectrophotometric–Kinetic Methods

The system described here permits acquisition of a large number of data with very accurate values of both absorbances and times. In addition, it simplifies data treatment and greatly facilitates application of kinetic methods to routine analysis. It has proved to be very convenient for the conventional initial rate, fixed time, and fixed absorbance methods.

Apparatus. The block diagram of the system is shown in Figure 8.13. A Commodore VIC-20 microcomputer provided with 16 kbytes of memory expansion (VIC-1111, a Commodore MPS 801 printer, a floppy disk unit and monitor, a Crison Digilab 517 potentiometer provided with an

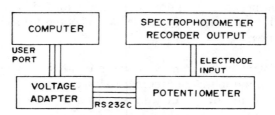

Figure 8.13. Block diagram of the data acquisition system for spectrophotometric–kinetic data.

RS232C interface, a TTL/RS232C voltage adapter, and a Beckman ACTA III spectrophotometer were used.

Software. The program KINSPEC was developed to acquire and treat the data. The main menu contains the following options:

1. *File Retrieving.* It allows data processing of previously filed experiments. First, it is verified if the computer contains data which have not yet been stored. In this case, the program asks if the current data must be saved. A positive response leads to the main menu.

2. *Experimental Task.* It starts the data acquisition process. A menu to set the experimental conditions is presented first. Parameters to be input are:

   (a) Kind of solution to be studied: a standard (its concentration is then asked for) or a sample.

   (b) Delay time from the beginning of the reaction until the first reading.

   (c) Reading frequency.

   (d) Total number of readings. Hereafter, the time count begins when any key is pressed.

   Data are iteratively acquired. If necessary, transmittance is converted to absorbance, and absorbances and times are stored. When the total number of readings have been acquired, the program returns to the main menu.

3. *Linear Regression.* Two possibilities are considered, that is, linear regression of absorbances versus time and of absorbances versus logarithm of time, which allows the linearization of some kinetic curves. The data are displayed and the limits of the interval to be fitted are required. The resulting slope, intercept, and intercept error are displayed and may be optionally printed. If the selected interval is not satisfactory, a new one may be selected. The data are stored and the program returns to the main menu.

4. *Data Printing.* It provides a hardcopy of the experimental data in table form.

5. *Calibration Curve.* It performs calibration curves from data previously acquired using a series of standards. First, the files obtained with the standards are retrieved; the type of file is verified and disregarded if it does not correspond to a standard. Second, intercepts and slopes of the normal and logarithmic regression straight lines are also retrieved. Finally, linear regression of the concentration–slope and/or concentration–log(time) points is performed; data are displayed and optionally printed and the calibration curve filed.

6. *Sample Processing.* The concentration of the analyte in the samples is calculated using the calibration curve.

7. *Data Saving.* An appropriate file is created, according to the kind of experimental task performed, that is, standards or samples.

8. *Exit to DOS.* The program verifies if there are unsaved data, in which case it returns to the main menu. Otherwise, it goes to DOS.

Results. The program KINSPEC was applied to the following kinetic procedures:

1. *Determination of traces of chromium(VI) with o-tolidine and H_2O_2.* In the presence of H_2O_2 and a trace of chromium(VI), *o*-tolidine (3,3'-dimethylbencidine) is oxidized into a blue compound with a maximum absorbance at 630 nm. Linear ranges were 0.1–2.5 μg/mL for the initial rate, fixed absorbance, and fixed time methods. The program was used for the determination of chromium in steel and tap water with errors below 2%.

2. *Determination of traces of manganese(II) with hydroxamic acid.* Hydroxamic acid reacts with manganese(II) to slowly form colored complexes. The applicability of KINSPEC to the study of complex equilibria was verified with terephthal-monohydroxamic acid with satisfactory results. Linear ranges were 1.2–7.5 μg/mL.

8.6. AUTOMATION OF FLUORIMETRIC METHODS

We developed the program package FLUOROPACK to facilitate experimental data acquisition in fluorimetric and phosphorimetric methods. The program has been used with an LS-5 Perkin-Elmer fluorimeter, which was optionally provided with an RS232C interface and connected to an IBM PC. The package also contains many data treatment and display techniques.

The structure of FLUOROPACK is shown in Figure 8.14. It was designed to meet the following criteria:

1. Modular structure. The user can select any of the different programs of the package from a main menu and come back to it at any moment.
2. Easy to operate. All programs are provided with an error trap to prevent any wrong input. An error message informs the user of the type of error and the number of the error line. A help subroutine is also available.
3. Easily extendable. Further modules can be added to the basic standard set of programs without difficulty.
4. Data storage in diskette and optional printer output.
5. Optional data manipulation, such as plotting, zooming, smoothing of spectra, and derivatives.
6. Compatibility. The package contains a conversion program for the data files to be made compatible with the classic package SURPHER (Golden Software Inc.).

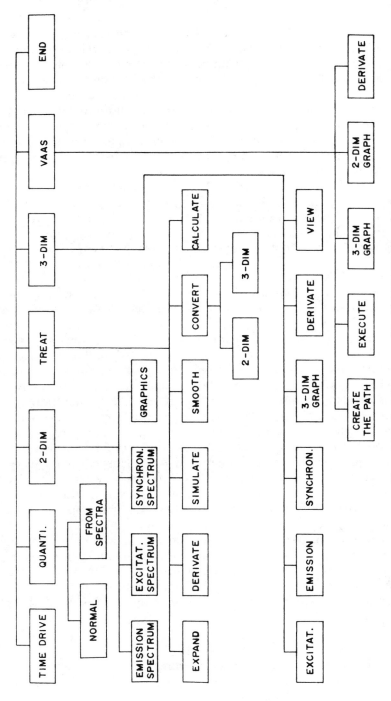

Figure 8.14. Overall structure of the FLUOROPACK package.

The Program VAAS of the FLUOROPACK Package. Fluorimetric techniques are generally considered as very selective. However, they have scarcely been used in multicomponent analysis of complex mixtures of fluorescent components. One reason is that the fluorescence spectra of individual substances consist of broad bands that may easily overlap.

Several methods have been proposed to solve this problem: among them synchronous and derivative fluorescence are the most popular. In these methods, the fluorescence intensity is presented as a function of a single parameter, which can be the excitation or the emission wavelength or the difference between them.

More information is obtained from a tridimensional spectrum in which the fluorescence intensity is a complex function of the excitation and emission wavelengths. Tridimensional data acquisition and treatment have been accomplished by a number of workers. However, despite the developments in instrumentation, the acquisition of a tridimensional data matrix and the necessary data reduction for further treatment are still very time-consuming processes.

From the three-dimensional spectrum, a contour map may be obtained, in which contour lines connect the points of equal fluorescence intensity. In this contour spectra (see Figure 8.15) a 0° section is equivalent to an excitation spectrum and a 90° section is equivalent to an emission spectrum. Synchronous (or fixed angle, with a fixed difference between both monochromators) spectra are represented by 45° sections.

To achieve still greater selectivity and, at the same time, to avoid the massive data acquisition and treatment processes of the tridimensional methods, variable angle fluorescence spectroscopy was developed. Variable angle fluorescence spectra are obtained by scanning both monochromators (excitation and emission) at different speeds, and thus the wavelength difference between them is no longer constant. Because the recorded spectra are not restricted to 45°, any path, even nonlinear, may be traced according to the particular sample under study. With variable angle fluorimetry, the simultaneous analysis of several compounds may be done from a single spectrum, without matrix reduction.

The program VAAS permits researchers to obtain variable angle fluorescence spectra without the need for hardware modification of the spectrofluorimeter. Instead, the program takes control of the motors that move the monochromators. This is much simpler and more flexible than implementing hardware modifications. The program VAAS performs the following tasks:

1. Creates a file with the necessary data to initiate scanning.
2. Executes the asynchronous scanning.
3. Obtains a tridimensional representation of the variable angle asynchronous spectrum.

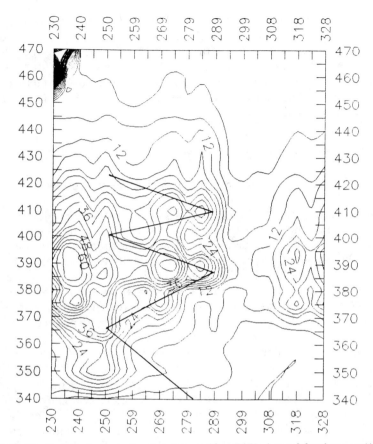

Figure 8.15. Contour fluorescent map of a mixture of four PAHs: benzo[a]anthracene (40 ng/mL), anthracene (8 ng/mL), phenanthrene (100 ng/mL), and pyrene (160 ng/mL). The straight lines show the route selected for the variable angle scan, which passes through emission maxima of the individual components.

On the other hand, information obtained from mixtures of fluorophores using VAAS may be treated with the program MULTIC. Assuming that, for a mixture, the fluorescence intensities are additive, the following general equation for each wavelength is deduced:

$$I_i = \Sigma k_{ji} C_j, \tag{8.1}$$

where I is the relative fluorescence intensity, k the sensitivity factor, and C the concentration; i corresponds to an excitation and emission wavelength pair and j to a fluorophore. For a complex mixture, the system of equations which corresponds to a variable angle spectrum may be solved by means of a multilinear regression method, using standards to first evaluate the sensitivity matrix. The program MULTIC was designed to set this fitting.

The program was tested with a mixture of fluorescein and dichlorofluorescein, mixtures of pesticides, and with a mixture of four polycyclic aromatic hydrocarbons (PAHs, see Figure 8.15) with satisfactory results. For maximum sensitivity, the variable angle route was traced through maxima of the individual spectra.

8.7. AUTOMATION OF CALORIMETRIC METHODS

8.7.1. A System for Automatic Thermometric Titrations

Apparatus. The block diagram of the system is shown in Figure 8.16. The power supply and Wheatstone bridge have been described previously (see Section 5.3.2). The thermistor (NTC, glass encapsulated thermometer type, with 100 kΩ) was submerged into a thermically isolated cell. A heat calibration system was also implemented using precision resistors and was employed both to determine the heat capacity of the cell and to rapidly reach the initial temperatures selected for the thermometric titrations. The thermometric cell consists of a polystyrene beaker suspended in a Dewar, which in its turn is surrounded by a porexpan block.

Other elements of the system are a 10x dc amplifier, a Crison Digilab 517 potentiometer, a Crison 738 autoburette, a paper chart recorder, an IBM PC with 256 kbytes of RAM memory, and a printer. The computer, potentiometer, and burette are provided with RS232C interfaces. The potentiometer is used as an A/D converter and as the RS232C interface for the signal. For this purpose, the Wheatstone bridge output is connected to the "electrode" input of the potentiometer through a 10x amplifier. Amplification takes advantage of the full scale range of the potentiometer, thus obtaining better precision in the A/D conversions.

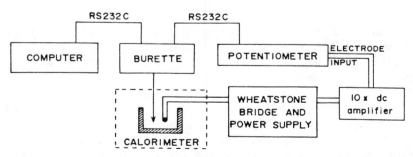

Figure 8.16. Block diagram of the system for automatic thermometric titrations.

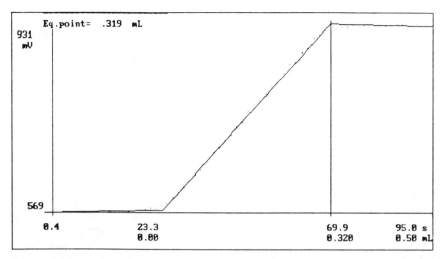

Figure 8.17. Thermometric titration curve of pyridine with HCl obtained using VALTER.

Software. Logic support of the system was provided by the program VALTER, which was designed to meet the usual requirements of menu driving, diskette storage, printer output, and data treatment facilities.

The experimental titration begins by filling up the syringe and by checking the temperature base line. The titrant is not added until a satisfactory temperature stability is achieved. For this purpose, the points taken during a previously selected time interval or "window" are linearly fitted. The slope of the regression straight line and the value of the correlation coefficient r are compared with the values initially selected by the user. When the requirements are met, the thermometric titration begins; otherwise the base line "window" is shifted by taking another temperature reading. The temperature curve is stored in the RAM memory until all the selected titrant volume has been injected.

Data treatment includes plotting of the normal, first and second derivative curves, linear fitting of intervals, calculation of the crossing points of the straight lines and, hence, of the uivalence point. An enthalpogram, which corresponds to the titration of pyridine with HCl in aqueous medium, is shown in Figure 8.17.

Programs for the evaluation of thermodynamic parameters (ΔH_i, $\log \beta_i$), such as MINITERM, can easily be applied if a word processor is used to adapt the data files. VALTER and MINITERM have been applied to the determination of the enthalpies of protonation and complex formation of several organic ligands of analytical interest.

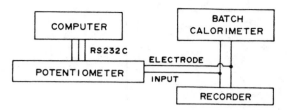

Figure 8.18. Block diagram of the batch calorimetric system.

8.7.2. Automation of a Batch Calorimeter

Thian-Calvet classic calorimeters are frequently used for the accurate evaluation of reaction enthalpies. The reagents are introduced in separated parts of the same cell, and when the system has reached thermal equilibrium, the cell is inverted to allow mixture of the reagents. The reaction heat is measured by a thermopile as it flows from the cell to the isothermal surroundings. The total heat evolved is obtained by integration of the resulting band-shaped signal. Manual integration is inaccurate, tedious, and time consuming. Therefore, we developed a data acquisition and processing system for this type of batch calorimeter.

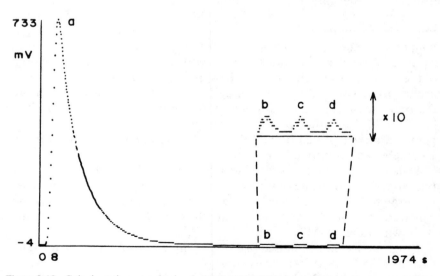

Figure 8.19. Calorimetric curve obtained with BATCHCAL: (a) peak due to the heat of formation of the Co(II)-5′-adenosine monophosphate complex; (b), (c), and (d), amplified peaks due to the friction heat.

Apparatus. The block diagram of the system is shown in Figure 8.18. A batch 2107 LKB microcalorimeter with an output of 1 V full scale, a Hetofrig CB7 cooling bath, an IBM PC computer provided with 256 kbytes of RAM memory, graphics capability, and an RS232C interface, a Crison Digilab 517 potentiometer with RS232C interface, a printer, and a recorder were used.

Software. The program BATCHCAL was written in BASIC and compiled. To perform an experimental task, the general conditions of the experiment, such as time between readings and maximum slope allowed for base line, are first requested and stored in a data file. The communication channel is opened and readings are taken until the base line slope is smaller than a prefixed value (10^{-5} mV/s in our experiments). Then the mixing process is manually started and data continue to be taken until the curve returns to the base line. Then, the user can either obtain a new calorimetric curve (e.g., to calculate the heat of friction) or finish the experimental task. The process can be repeated a number of times. When finishing, the data are filed.

The curves and data sets can be displayed on the screen, the peak integrated, and the data and final results printed. A calorimetric curve obtained with BATCHCAL is given in Figure 8.19.

8.8.   AUTOMATION OF FLOW INJECTION ANALYSIS (FIA)

8.8.1.   Introduction

The automation of most analytical procedures is made easier by the use of flow systems. Segmented and unsegmented systems are usually distinguished. Unsegmented systems have developed very rapidly since the introduction of the flow injection analysis (FIA) technique in the mid-1970s. Advantages of FIA are high sampling rate, low consumption of sample and reagents, simplicity, versatility, and low cost. Consequently, FIA techniques have replaced segmented systems in many applications, with the exception of complex analysis procedures requiring the determination of many parameters of each sample.

Automation of flow injection techniques is particularly attractive due to the wide field of potential applications. To develop totally automated FIA systems the following elements are necessary:

1. One or two multichannel peristaltic pumps with standard interfaces.
2. An injection valve with pneumatic actuator and electrical controller.
3. A programmable sampler.

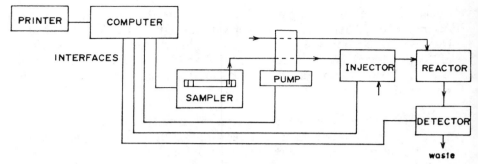

Figure 8.20. Scheme of a basic, completely automated, FIA setup.

4. Several detector moduli (optical, electrical, etc.) with analog or stand-ard digital output and input facilities, the latter for the automatic control of experimental conditions (wavelengths, potential, gain, etc.).

The scheme of a basic, completely automated, FIA setup is shown in Figure 8.20.

8.8.2. An Automatic FIA System for the Simultaneous Determination of Nitrates and Nitrites in Waters

Apparatus and Manifold. A scheme of the FIA setup and data acquisition system is shown in Figure 8.21. The system consists of a Gilson Minipuls 2 peristaltic pump (8 channels), a Rheodyne 50 injection valve, a thermo-static water bath, an ACTA III double-beam spectrophotometer provided with flow-through cells of 18 μL and 1.000 cm optical path, a Crison Digilab 501 potentiometer with RS232C interface, and an Amstrad CPC6128 microcomputer provided with RS232C interface and a floppy disk unit. All reactor coils and sample loop tubes were made of PTFE (0.5 mm i.d.). Data acquisition was implemented by connecting the recorder output of the spectrophotometer to the potentiometer.

Procedure. The method is based on the classic diazotization and coupling reactions for nitrites. The sample is split into two flows:

1. One of them is directly treated with the azodye-forming reagents and sent to the sample flow cell of a double-beam spectrophotometer, which allows determination of the nitrite contents.
2. The other flow goes through a copperized cadmium microcolumn, where nitrates are reduced to nitrites. The sample is then treated with

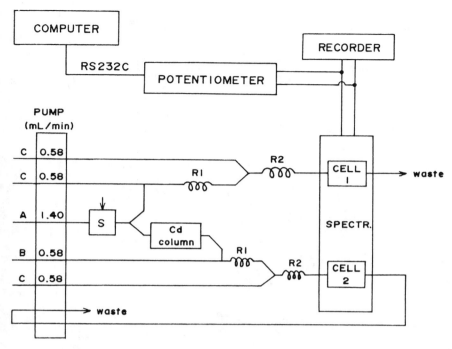

Figure 8.21. Scheme of the FIA setup and data acquisition system for the simultaneous colorimetric determination of nitrates and nitrites: A = carrier, B = sulfanilamide, C = *N*-(1-naphthyl)ethylenediammonium chloride, S = injection valve, R1 = 10 cm reaction coil, R2 = 30 cm reaction coil.

the reagents to form the azodye and the overall mixture is sent to the reference cell of the same double-beam spectrophotometer. The absorbance gives the sum of nitrates and nitrites.

Software. A program was developed with the following options: (1) data acquisition and storage of calibration curves, (2) data acquisition and storage for unknown samples, (3) calculation of the calibration curves, and (4) output of reports of the nitrate and nitrite contents.

When option 1 is selected, the screen is divided into three windows: in the first (top left), the current FIAgram is represented, which monitors the validity of the acquired data; in the second (bottom left), an optional menu and messages are displayed; and in the third (right half screen) the data are displayed.

The analyst may introduce the successive standards in any order. Once all standards have been introduced, the computer plots the data and the calibration curves. On menu, outliers may be removed, additional stand-

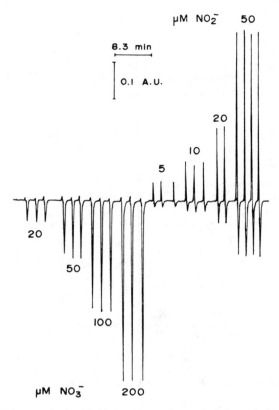

Figure 8.22. FIAgram obtained in the presence of several nitrite and nitrate concentrations.

ards introduced, or the overall process aborted. The calibration curve is stored and optionally printed.

When measurements of unknown samples are taken, the monitor screen is organized in a similar way as for the calibration subroutine. The maximum and minimum of each peak are displayed in the window for results, together with the concentrations computed for nitrates and nitrites. The acquired data may be saved on a floppy disk. If the data are saved during the process, any desired number of injections for each sample may be carried out. However, only four measurements for each sample are displayed simultaneously on the screen. Optionally, a hardcopy of the results may be printed for each working session.

Results. When nitrates were injected, negative peaks (sample flow cell) were obtained, whereas nitrites gave both negative (sample flow cell) and positive peaks (reference flow cell). A FIAgram is shown in Figure 8.22.

Three calibration curves can be obtained—one for the negative nitrate + nitrite peaks, the second one for the nitrite negative peaks, and the third one for the nitrite positive peaks. Nitrite concentration can be obtained directly from the third curve. In order to calculate the nitrate contents, the negative signal corresponding to nitrites had to be first calculated from the second calibration curve and subtracted from the overall negative peaks of the first calibration curve. The linear dynamic ranges were 10–200 µg/mL for nitrates and 2.5–50 µg/mL for nitrites. Relative standard deviations ranged from 2 to 12%.

CHAPTER

9

LABORATORY ROBOTICS

9.1. INTRODUCTION

In comparison with the intelligent skilled robots of the science-fiction literature, laboratory robots of today look like their rudimentary ancestors. They are usually reduced to an articulated arm surrounded by the instruments and devices that make it possible for the arm to perform a process automatically. Thus, for instance, the homemade sampler shown in Chapter 7, with an arm provided with a single movement, may be considered a simple robot. This and other types of laboratory robot, consisting of a computer or controller and a more or less complex articulated arm, permit very high levels of automation. Articulated arms make it possible to automate the more troublesome steps of many analytical procedures, such as sampling, sample milling/reduction cycles, weighing, solving, sample conditioning, and injection (see Table 1.1).

Versatility is the great advantage of robots when compared with other automatic mechanical devices, since a robot can be reprogrammed to perform many different tasks. The introduction of robots represented a revolution in mechanics, which in some way was parallel to the impact produced by the microprocessor in electronics. Robots, like microprocessors, can attain a high level of flexibility and excellent performance on the basis of the substitution of software by hardware.

With the development and popularization of microcomputers in the late 1970s, a parallel development of robotics was expected. However, the introduction of robots in laboratories, as well as in many other scientific and industrial environments, has been very slow. Consequently, the research projects of several manufacturers related with laboratory robots have been delayed or cancelled.

This is a consequence of the higher prices of robots. The potential buyers of robots on a large scale are microcomputer users; but no truly capable robot arm has been made available at a price that micro users can afford. Today, most laboratory robots are found in situations where they are really needed, such as in risky environments (e.g., working with radioactive substances) or when valuable substances are handled (e.g., some pharmaceuticals and fine chemicals). The popularization of robots will probably come about gradually, as the slow-moving mechanics industry becomes capable of providing the appropriate parts at low cost.

Advantage of robots in the laboratory are in part coincident with the advantages of computers and may be summarized as follows:

1. Easy programming by software rather than by hardware.
2. Improved reproducibility and reliability (since robots are tireless).
3. Increased profitability of equipment, which can be kept working day and night without stopping.
4. Lower operation costs when a large volume of work must be done (today, robots are expensive).
5. Ability to work in risky environments.
6. Possibility of automating complex operations, such as sample conditioning.
7. Easy automation of traditional manual procedures, due to the human-like behavior of robots.

Some basic devices in robotics, used to construct programmable mechanical systems, are described next.

9.2. MOTORS

9.2.1. Introduction

Frequently, in an experiment the position of objects must be changed. In some cases, such as when a valve is adjusted or a micrometer is turned, the action is inherently rotary. In others, the object must be moved linearly, which is accomplished with a screw type drive (leadscrew) so that it may be considered rotary. Generally, positioning requires the use of motors as output transducers of rotary movement.

The final objective of the rotary action may be either to continuously drive an object, such as a light chopper, or to move accurately into position an object, such as a sample holder. In the former case, accurate speed control is required, while in the latter, position must be accurately sensed.

9.2.2. Synchronous Motors

The speed of synchronous motors is determined by the power line frequency. They are useful where a rather slow but reasonably constant speed is required. The speed is fixed at manufacture and can only be changed by changing the line frequency. Synchronous motors should be considered unidirectional, although change in direction is possible for some by using an appropriate phase-shifting circuit.

9.2.3. Direct Current Motors

Direct current motors furnish straightforward control of direction and speed. The direction is determined by the polarity, and speed by the applied power. They come in all sizes and varieties, from miniature high-speed motors to large powerful motors with a wide speed range.

The speed of a dc motor is related to the power consumed, which is approximately proportional to the time integral of the voltage. Two approaches follow for speed control: variation of the dc voltage and pulse-width modulation. In either case, the speed is only controlled in a relative sense; the applied voltage, current, and torque interact nonlinearly and these relationships depend on the type of motor. A separate tachometer is required to absolutely determine the speed.

The dc voltage can be continuously controlled by using a D/A converter. The provision of sufficient current may require an adequate power amplifier. The motor can be stopped quickly with a circuit that reverses momentarily the polarity of the power.

9.2.4. Servo Motors

Servo motors are parts of positioning systems designed on the principle of feedback. A drive signal is sent to the servo control system representing the desired position. The actual position is continuously measured by a transducer, such as a potentiometer configured as a voltage divider, and it is fed back to the servo control system which develops the difference or error signal. That error signal is amplified to activate the motor. As the motor moves, the error signal moves toward zero, and when the difference between drive signal and feedback signal is zero, the motor stops.

Usually, the feedback loop is closed in hardware: the drive voltage is compared with the output of the potentiometer that encodes the desired position. The loop can also be closed by software: the computer reads the position with a potentiometer and an A/D converter. The difference between that value and the desired position determines whether the computer should turn on the motor.

9.2.5. Stepper Motors

As long as high speed is not required, stepper motors are ideal translation devices to be placed under computer control. The stepper motor's principle of operation relies on the response of a multiple rotor to a precisely pulsed voltage pattern applied to the surrounding stator windings. The shaft of the stepper motor responds by rotating an exact fraction of a turn, usually given in step angles, for each sequential voltage pulse applied to the motor windings. The discrete positions are fixed by the position of coils within the motor. Each step of rotation is commanded separately, a move from one position to the next, so complete control over position is obtained.

The error in moving a position is not accumulative, but limited to the uncertainty in the last single step. Maintaining the position information is simply a matter of keeping track of the number of steps that have been performed. No feedback mechanism for speed or position is required as long as the motor is able to respond to the step commands. Consequently, it is easily placed under digital control.

Stepper motors are very convenient devices because their steps are precise, uniform, and reproducible. Frequently, they are the choice for robots and other precision interactive-control applications, such as stirrers with reproducible rpm values, printers, plotters, recorders, clocks, and disk drivers.

The operation of a stepper motor can be understood by referring to Figure 9.1, which shows a scheme of a permanent magnet stepper motor with only four phases per revolution. The stationary component, or stator, consists of four separate windings, while a permanent magnet forms the rotating component or rotor. The rotor, moving in discrete steps, will seek

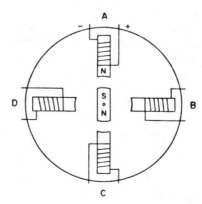

Figure 9.1. Scheme of a simple stepper motor with four phases per revolution.

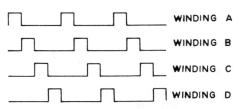

Figure 9.2. Excitation wave of the stepper motor of Figure 9.1.

the position that experiences the strongest pull. Since each stator winding is independent, that position is precisely determined.

In this simple case, the windings are sequentially energized in wave excitation sequence states (Figure 9.2). Each winding develops a north pole facing the permanent magnet, and the south pole of the rotor follows the sequence. The rate of the switching sequence determines the speed, and the direction of the switching sequence determines the direction of rotation. A simple circuit for generating the logic state sequence may be used.

Energizing only a single phase at a time makes inefficient use of the windings when power and speed must be maximized. In the most common approach, improvement is gained by energizing two adjacent windings simultaneously in a two-phase excitation technique.

If there are only four stator windings per revolution, the motor will make

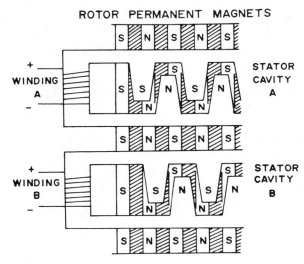

Figure 9.3. Scheme of the stator windings and permanent magnet rotor of a stepper motor.

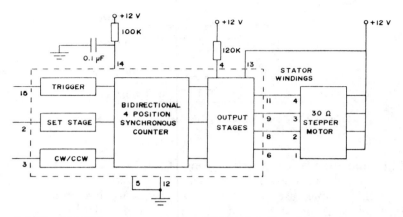

Figure 9.4. The SAA1027 stepper motor driver IC. Pins 1, 7, 10, and 16 are not connected.

large, full steps of 90°. The resolution is increased by employing repeated sequences of four phases, as shown in Figure 9.3. In practice, the largest step size for commercial stepper motors is 15°, which corresponds to 24 steps or sets of windings per revolution. Stepper motors with 7.5° per step (48 steps per revolution) are also common. Motors having step angles as small as 0.72°, that is, 500 steps per revolution, are also available.

A combination of a power amplifier and the logic circuit to drive a stepper motor is available in an integrated circuit. For instance, the SAA1027 is intended for driving a four-phase two-stator stepper motor. As shown in Figure 9.4, the circuit consists of four output stages and a logic part that is driven by three input stages: a trigger input stage, an input stage that can change the switching sequence so that the motor can rotate clockwise (CW) or counterclockwise (CCW), and a set input stage to set the four output stages. The three inputs are compatible with high noise immunity logic to ensure proper operation even in noisy environments. The output can deliver 350 mA in each phase. The right switching sequence of the four phases is obtained from the logic part of the circuit. Integrated diodes protect the outputs against transient spikes. The supply voltage is 9.5–18 V.

9.3. POSITION DETECTORS

9.3.1. Introduction

The need to test the position or proximity of a component in an experiment is a common requirement. The problem may be as simple as determining the on or off position of a switch, or it may involve synchronization of an event with the phase of a high-speed rotating device. Complete control over the physical translation of an object includes both position of that object and the rate at which the object moves.

Most dc and ac motors are capable of accurate movements, even at very high speeds, when separate transducers feed back information about position. Sometimes a combination of dc and stepper motors is useful, the former for speed and the latter for accuracy.

There are many transducers capable of proximity or position detection. Some are inherently digital and are capable of only two signal levels. Others are basically analog, and a threshold must be set to discriminate between two levels. Some common position transducers are described below.

9.3.2. Mechanical Switch Closures

Mechanical switches are inherently binary, since the contact of the switch is either closed (making contact) or open. Mechanical switches serve several needs. The first is as an interface to the operator. If a panel or keyboard switch is used by the operator to enter information (data, reset, etc.), the computer must read the position of that switch. Mechanical switches are also used as part of the interface to an experimental device. Small precision switches, often called microswitches, are used to indicate with high reproducibility whether a component is or is not in position.

A microswitch placed at the end of a leadscrew carriage can be used as a safety to prohibit further travel. A microswitch can also provide a means of calibrating the position of the carriage, its closure indicating that the carriage has reached a known position. This is very useful, for example, for the initialization of a system. Examples of these two uses of microswitches may be found in the homemade autosampler, described in Section 7.8.2.

When a switch is closed, the corresponding order of stopping the movement can be provided by hardware, with the appropriate circuit, or by software. In the latter case, the event may be read by a computer as a high in a digital input of an interface card. To convert the mechanical switch closure and opening into two digital TTL levels, the circuit of Figure 9.5

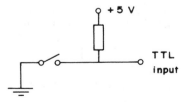

Figure 9.5. Circuit to provide the two TTL levels with a microswitch.

may be used. If one terminal of the microswitch is connected to ground and the other to the TTL input, closure grounds the input, thereby providing a low; but when the switch is open, no current can flow to ground and the input is at 5 V, thus giving a high.

9.3.3. Noncontact Position Detectors

In order to operate the mechanical microswitch, physical contact between the object and the switch must be made. This may be undesirable for any of several reasons: the switch must be outside the container, the environment might damage the exposed contacts, or the moving object might be affected by striking the switch.

By coupling a phototransistor (PT) with a small light source, such as an LED or a small He–Ne or diode laser, a position may be tested by determining whether the light beam is or is not blocked. Assemblies are readily available that include both an LED as a light source and a transistor as a detector, both integrated into a plastic housing for easy mounting (see Figure 9.6). These devices come in a variety of shapes and are known as optointerrupters. They are functionally the same as the optocouples, except for the addition of an accessible gap in which the light path can be interrupted.

A variant can be used to monitor the presence of an object capable of reflecting the light. Crison uses this system to detect the position of the samples and the home position of the Model 2040 microSAMPLER. Reflective strips at the sample positions encode the sample number (see Figure 9.7). When the strip is located in front of the detector, the light from an LED is reflected and reaches the base of a phototransistor, which is brought into conduction.

9.3.4. Potentiometers

Potentiometers are used for the continuous measurement of position. Potentiometers are passive analog position transducers. Wired as a voltage

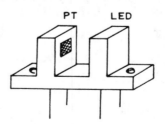

Figure 9.6. A LED–PT couple.

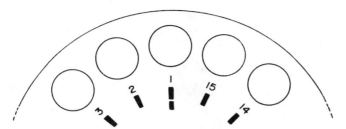

Figure 9.7. Reflective strips for detecting position in the carousel of the Crison 2040 microSAMPLER.

divider, the potentiometer voltage at the wiper is proportional to the position between the extremes. Precision potentiometers are manufactured in several varieties. The total excursion may be less than one full turn or up to 30 turns. Single-turn potentiometers without stops at the extremes can also be found.

The immediacy of the output of potentiometers has earned them a successful application in servo devices, mainly in strip-chart recorders. Potentiometers are also used to feed back software-controlled systems. A useful example is provided by the Fischertechnik interface. As shown in Figure 5.27, the interface has two lines to digitalize resistance measurements. If the resistor is a potentiometer, the resistance can be related with the position of the wiper. This is used in a Fischertechnik prototype robot for the accurate detection of the turn displacement with respect to the vertical position. Since the precision of the A/D converter is 1/255, the turn position is detected with a precision of about 1°.

9.3.5. Shaft Encoders

Incremental shaft encoders are digital devices. Pulses are generated when the rotation of a ruled or slotted disk is sensed by optical interrupters (see Figure 9.8). These pulses are counted discretely. Then, with appropriate circuitry, an up–down counter can maintain a record of the absolute position. Thus, in the Fischertechnik prototype robot, shaft encoders are used to control several movements of the articulated arm.

Similarly, linear encoders are used to control linear shift movements (see Figure 9.9). The disk shaft encoder will usually be preferred, because much greater resolution can be achieved. Applications of the linear encoder are limited to cases in which no conversion from linear position to rotation is available.

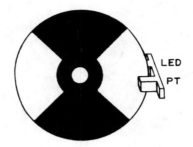

Figure 9.8. A low-resolution shaft encoder.

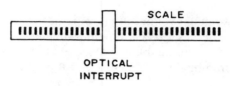

Figure 9.9. A linear encoder.

9.3.6. Tachometers

Tachometers provide information about the rate of rotational displacement. Direct current tachometers are electrical generators from which a signal is drawn. Since no power is required, the windings can be very light in order to reduce friction.

Usually, a dc tachometer is mounted on the same shaft as the motor that

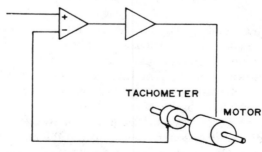

Figure 9.10. Control of the speed of a motor with a tachometer in the feedback loop of an OA.

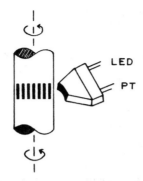

Figure 9.11. Optical tachometer.

actuates the rotation. Very precise speed control is possible when this type of tachometer is used in the feedback loop of an operational amplifier that drives the motor (see Figure 9.10). Alternatively, the output can be fed to the input of an A/D converter for software control.

A very useful variant is the optical tachometer, in which an optical encoder employs an optical interruptor in a reflective mode (see Figure 9.11). An LED–PT couple measures the rotating speed by means of a reflective surface, such as a piece of reflective tape, on a rotating shaft. In order to accurately and precisely convert the resultant pulses to a rate, the pulses can be counted for a preset period either with a hardware counter or by using interrupts and a software counter. If instantaneous control of rotational speed is desired, the pulses can be converted to a dc voltage as part of a feedback loop.

9.4. ARTICULATED ARMS

9.4.1. Introduction

There are several commercial articulated arms designed for laboratory tasks. Some manufacturers are Zymark, Perkin-Elmer, HP Gemchem, UMI (Universal Machine Intelligence), and LABMAN.

In the Zymate robotized lab system from Zymark, the arm parts move around cylindrical coordinate axes, following the simple movements depicted in Figure 9.12: turn around the main vertical axis (movement 1), up and down (2), forward and backward (3), hand roll (4), and opening–closing of the gripper (5).

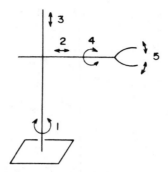

Figure 9.12. Movements of the Zymate robot arm from Zymark.

In contrast, the Perkin-Elmer robot has an anthropomorphic arm with spherical coordinates (Figure 9.13). Its movements are more complex and resemble more closely those of the human arm. The arm turns 360° around the main vertical axis and performs movements similar to those of shoulder, elbow, wrist, and hand. This configuration is the most frequently found among commercial arms. Some robots have a hybrid design, such as the Model RTX from UMI.

9.4.2.   Features of Articulated Arms

Articulated arms are characterized by a number of parameters which are described next.

Drive Systems. The drive systems provide the mechanical power for the movements of the arm parts. They can be pneumatic, hydraulic, ser-

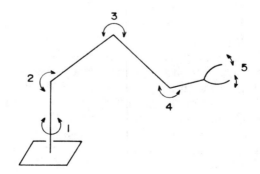

Figure 9.13. Movements of the Perkin-Elmer robot arm.

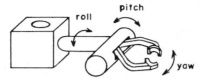

Figure 9.14. The three movements of the hand of the RTX robot from UMI.

vomotors, or stepper motors. Pneumatic and hydraulic systems are preferred in industrial applications since they provide speed and strength. Servomotors and stepper motors are more adequate in laboratory robots. To achieve the required accuracy and reproducibility of the spatial (three-dimensional) positions and movements, optical devices such as shaft encoders are used.

Hands. These are important accessories, since the design of the hands and their possible movements limit the operations that the robot is able to perform. A feature to be considered in the design of the hands is the number of axes of rotation. Robots are frequently characterized by this number. Some special movements that are usually not considered in industrial robots, such as the wrist turn, are very important in laboratory robots. The three movements of the hand of the RTX robot are shown in Figure 9.14.

An interesting feature of a laboratory robot is the possibility of automatically changing the hands from a hand parking station located at the robot reach (see Figure 9.15). This makes it easier to adapt the robot to different tasks (e.g., weighing, pipetting, injecting into a chromatograph) and allows higher degrees of efficiency and precision to be reached. These changes of task are frequently necessary within any analytical process.

Sensors. The number and type of available sensors are also important features of a robot. Sensors are used to feedback robot operation, thus avoiding errors. For instance, a regulable pressure sensor may be used to detect if the robot has effectively grasped an object and to control the pressure on it. This is very important when valuable, toxic, or corrosive substances are handled. However, in any case, an undetected error can trigger a sequence of catastrophic events.

Robots, just like any other automatic system, are absolutely stupid and incapable of making any decision. Care must be taken in programming a robot to anticipate all possible situations, paying particular attention to the most obvious and elemental problems that we solve instinctively.

Sophisticated robots are provided with optical sensors, such as photoelectric cells, laser vision (by reflection of a laser beam), and CCD cameras.

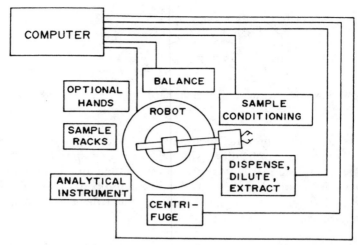

Figure 9.15. Scheme of the robot lab system from Zymark, which includes a hand parking station.

Frequently, commercial robots include digital and analog I/O devices that can be optionally added to the robot to connect user's sensors.

Training Programs. Programming of an articulated arm to perform all the steps of a given process can be done using the following two methods:

The most obvious and crude way is to calculate the coordinates of all axes of movement for each desired position of the arm, and to introduce all the resulting values in a file to be sequentially read by the computer.

The other method is to use a training program. This is a very common, convenient, and less risky method than the first one to program a robot. With a training program, the user gives instructions to the robot with the objective of performing the entire sequence of desired movements, like in a rehearsal session. Each new position is tried, as many times as necessary. The data can be input through the computer console, through an additional keyboard, or with a joystick. The coordinates of all the axes are continuously monitored by the computer, whether counting steps of the motors or using information from optical encoders. When the user decides that the new coordinates are adequate, they are stored in a file. Thus, a program that contains the entire sequence of movements is created. When the training process finishes, this program can be used to execute the sequence as many times as desired.

**Table 9.1. Some Compatible Accessories of the
Zymate Laboratory Robot from Zymark**

PySection	PySectors (Occupied Sectors)
General purpose hand	2
Vibrating general purpose hand (for efficient transfer of powders and tablets)	2
Racks	4–5
Pipetting and filtering	3
Weighing (e.g., with Mettler AE balances)	5
Dilute and dissolve	3
Membrane filtration	2
Liquid/solid extraction	2
Liquid/liquid extraction	3
Linear shaking	8
Centrifugation	8
Evaporation station	4
Screw capping	3
Crimp capping (for GC or HPLC autosampler)	4
Spectrophotometer sip	2
LC inject	3
GC inject (e.g., with the Hewlett-Packard 7673A autoinjector)	8
Karl Fischer titration (e.g., with the Mettler titrator)	4
Master laboratory station (liquid handling using three syringes)	0
Power and event controller (PEC) (interfacing with other laboratory equipment)	0

The availability of easy-to-operate, powerful training programs is an important feature to be considered before buying a robot.

Environment. Together with the performance of the arm itself, the available compatible accessories, which constitute the environment of the articulated arm, must also be considered. The number and type of the accessories will limit the number of possible laboratory operations that the robot will be able to perform. Some accessories available from Zymark are listed in Table 9.1.

In a common design, the accessories are located in a circular way, like the sections of a pie, within the reach of the robot arm, which stands in the center. However, in some cases, as in the Perkin-Elmer or the LABMAN robots, the articulated arm can also move on a pair of tracks, to reach any place on a bench top.

9.5. THE TRAINING PROGRAM OF THE FISCHERTECHNIK PROTOTYPE ROBOT

As an application example let us see how the Fischertechnik robot training program works. The main menu is shown in Table 9.2. Option T serves to teach a new process to the robot, and the other options handle files that may contain other processes. When option T is selected, the menu of Table 9.3 appears on the monitor screen. This menu is used iteratively to create a sequence of robot arm movements that are stored in an open file. In fact, the conditions given in Table 9.3 correspond to the movement number 18 of a process.

In the Fischertechnik prototype robot, control of the position of the moving parts is achieved by means of shaft encoders. The related information is displayed in the teach-in mode menu in front of the F1–F4 lines. The corresponding function keys are used to change the arm position by changing the codes. The type and direction of the movement are established using keys 1–8. In the last line of the menu, the status of the new position of the arm appears. When the position is satisfactory, the corresponding data can be added to the file by pressing the return key. The final listing of a process may have the look shown in Table 9.4.

9.6. LABORATORY UNIT OPERATIONS

The modular concept of laboratory robots has led to the concept of laboratory unit operation (LUO). In a LUO only one device or instrument, located in a

Table 9.2. Main Menu of the Training Program of the Fischertechnik Robot

T	Teach-in mode
R	Execute program
S	Save program
L	Load program
D	Directory
P	Print program
E	End of program

Table 9.3. Teach-in Mode Menu of the Training Program of the Fischertechnik Robot

Key	Function	Position[a]
1	Robot ccw	4
2	Robot cw	
3	Upper arm forward	0
4	Upper arm backward	
5	Forearm down	0
6	Forearm up	
7	Open gripper	Closed
8	Close gripper	

Number of steps		4
Operation time gripper		200

F1[b]	256 Sectors	M	Menu[c]
F2	64 Sectors	Home	Home
F3	16 Sectors	Return	Teach
F4	4 Sectors	<−	Delete

Number	Body	Upper arm	Forearm	Gripper
18	0	0	0	Open

[a] Number of sectors (counts) given by the respective shaft encoder.
[b] From F1 to F4, function keys and associated number of preselected shaft encoder sectors to program the motors.
[c] From Menu to Delete, functions associated to the M, Home, Return and Backspace keys.

fixed place and with a given function, such as dispensing, pipetting, or mixing, is involved. LUOs can be arranged in a number of combinations, which makes it easier to design and program many automatic processes. In the PyTechnology, introduced by Zymark, each LUO is performed by a

Table 9.4. Example with First Eight Movements of a Robot Arm Process

Number	Body	Upper Arm	Forearm	Gripper
0	0	539	1072	Open
1	2	927	1072	Open
2	2	927	1072	Closed
3	2	694	1072	Closed
4	838	694	1072	Closed
5	838	962	1072	Closed
6	838	962	1072	Open
7	854	435	1072	Open
8	854	970	1072	Open

PySection. PySections consist of the appropriate hardware and software necessary for automating a particular LUO (see Table 9.1).

The Zymark Zymate robot is mounted on a central plate, around which all the necessary PySections are located. The central plate is divided into 48 sectors (7.5° per sector). The hardware for each PySection, such as a dispenser or centrifuge, is mounted on a wedged-shaped platform, which occupies one or more of the 48 sectors. The user informs the controller of the sector location of each PySection. This information is stored by the controller. All working positions for that PySection are adjusted to the stated location. No additional robot teaching or module programming is required by the user. This speeds both initial system setup and system reconfiguration when changing applications.

Thus, for instance, a PySection may contain a sample rack with the coordinates of its 50 vial positions. It will be enough to inform the controller of where the PySection has been relocated, for the 50 coordinates to be perfectly adjusted to the new positions of the vials. A complex and tedious training session of the robot is thus avoided.

The software included with the PySections is designed to make their installation and use simple and reliable. This software provides prenamed commands, which form a high-level application language for the implementation of each section. Prenamed commands contain the positional information and the best current techniques to perform each LUO. Thus, complex movements of the robot arm can be programmed in an easy way, as shown in the following examples:

GET.TUBE.FROM.RACK.n
PUT.TUBE.IN.FILTRATION.STATION
GET.1ML.TIP
ASPIRATE.1ML.TIP
DISPENSE.1ML.TIP

where each apparently simple instruction represents large sequences of coordinates and steps of motors.

RECOMMENDED BIBLIOGRAPHY

General

J. J. Brophy, *Electrónica Fundamental para Científicos*. Reverté, Barcelona, 1969.
J. J. Carr, *Data Acquisition and Control. Microcomputer Applications for Scientists and Engineers*. TAB Professional and Reference Books, Blue Ridge Summit, PA, 1988.
P. Horowitz and W. Hill, *The Art of Electronics*. Cambridge University Press, Cambridge, 1980.
K. L. Ratzlaff, *Introduction to Computer-Assisted Experimentation*. Wiley, New York, 1987.
B. H. Vassos and G. W. Ewing, *Analog and Digital Electronics for Scientists*. Wiley, New York, 1985.

Analog Electronics

M. Torres, *Circuitos Integrados Lineales*. Paraninfo, Madrid, 1984.

Digital Electronics

J. M. Angulo, *Electrónica Digital Moderna*. Paraninfo, Madrid, 1988.
P. R. Rony, *El Microprocesador 8080 y sus Interfases*. Paraninfo, Madrid, 1984.

Microprocessors and Microcomputers

J. M. Angulo, *Microprocesadores*. Paraninfo, Madrid, 1981.
J. W. Coffron, *The VIC-20 Connection*. Sybex, Berkeley, CA, 1983.
P. De Miguel and J. M. Angulo, *Arquitectura de Computadores*. Paraninfo, Madrid, 1987.
L. A. Leventhal, *8080A/8085 Assembly Language Programming*. Osborne/McGraw-Hill, Berkeley, CA, 1978.
Intel, *SDK-85 User's Manual*. Santa Clara, CA, 1978.
Intel, *Introduction to the 80386*. Anaya Multimedia, Madrid, 1989.
S. P. Morse and D. J. Albert, *The 80286 Architecture*. Wiley, New York, 1986.

M. Robin and Th. Maurin, *Interconexión de Microprocesadores.* Paraninfo, Madrid, 1982.

R. R. Smardzewski, *Microprocessor Programming and Applications for Scientists and Engineers.* Elsevier, Amsterdam, 1984.

R. Zaks, *Microprocessors from Chips to Systems.* Sybex, Paris, 1979.

R. Zaks, *Applications du 6502.* Sybex, Paris, 1980.

Communications

Capital Equipment Corporation, *P<=>488 Programming and Reference Manual.* Capital Equipment Corporation, Burlington, MA, 1986.

F. J. Ceballos, *Manual para QuickBASIC 4.0.* RA-MA, Madrid, 1988.

Hewlett-Packard, *HP-IB Command Library for MS-DOS,* Publication No. 82990-90001, 1986.

M. D. Seyer, *The IBM PC/XT—Making the Right Connections.* Prentice-Hall, Englewood Cliffs, NJ, 1985.

Interfaces

Data Translation, *User Manual for DT2811,* Data Translation Document UM-06160-B-2907. Data Translation, Malboro, MA, 1987.

Fischertechnik, *Interface IBM Personal Computer.* Fischertechnik, Fairfield, NJ, 1986.

MetraByte, *DASH-8 Manual.* MetraByte, Taunton, MA, 1984.

R. A. Penfold, *Técnicas y Proyectos de Interfases.* Anaya, Madrid, 1986.

Modular Elements

Hewlett-Packard, *HP 8452 Diode Array Spectrophotometer Handbook,* Publication No. 08452-90001. Hewlett-Packard, Palo Alto, CA, 1986.

Hewlett-Packard, *HP 8452 Diode Array Spectrophotometer. HP 89530 MS-DOS UV/vis Software Handbook,* Publication No. 89530-90002. Hewlett-Packard, Palo Alto, CA, 1986.

Hewlett-Packard, *HP 8452A Diode Array Spectrophotometer. Interfacing and Programming Guide,* Publication No. 08452-90005. Hewlett-Packard, Palo Alto, CA, 1986.

Hewlett-Packard, *HP 8452A UV/vis Software. Source Code.* Hewlett-Packard, Palo Alto, CA, 1986.

Keithley, *Model 175 Instruction Manual,* Document Number 175-901-01C. Keithley, Cleveland, OH, 1984.

Keithley, *Model 1753 Instruction Manual,* Document Number 1753-901-01C. Keithley, Cleveland, OH, 1984.

Laboratory Automation

J. K. Foreman and P. B. Stockwell, *Automatic Chemical Analysis*. Ellis Horwood/ Wiley, Chichester, 1975.

Robotics

Fischertechnik, *Programming/Kit-Building Instructions*. Fischertechnik, Fairfield, NJ, 1986.

Zymark, *PyTechnology Zymate Laboratory Automation Systems*, Zymark Corporation, Hopkinton, MA, 1986.

INDEX

B.C.

(*continued from front*)

RETURN CHEMISTRY LIBRARY
TO ➡ 100 Hildebrand Hall

JAN 14 1991

(*continued on back*)